Yosr Kadri
Ons Haddad
Khouloud Ben Youssef

The Universe of Bacterial Infections in Nephrology

Yosr Kadri
Ons Haddad
Khouloud Ben Youssef

The Universe of Bacterial Infections in Nephrology

ScienciaScripts

Imprint

Any brand names and product names mentioned in this book are subject to trademark, brand or patent protection and are trademarks or registered trademarks of their respective holders. The use of brand names, product names, common names, trade names, product descriptions etc. even without a particular marking in this work is in no way to be construed to mean that such names may be regarded as unrestricted in respect of trademark and brand protection legislation and could thus be used by anyone.

Cover image: www.ingimage.com

This book is a translation from the original published under ISBN 978-620-6-72446-9.

Publisher:
Sciencia Scripts
is a trademark of
Dodo Books Indian Ocean Ltd. and OmniScriptum S.R.L publishing group

120 High Road, East Finchley, London, N2 9ED, United Kingdom
Str. Armeneasca 28/1, office 1, Chisinau MD-2012, Republic of Moldova, Europe
Printed at: see last page
ISBN: 978-620-8-31847-5

Contents

1 INTRODUCTION

oday, chronic kidney disease (CKD) is considered to be one of the most serious diseases in the world.

Chronic kidney disease is a real health problem in Tunisia. In 2008, the prevalence of chronic kidney failure was 734 per million inhabitants. This figure has been rising steadily over the years [1]. Patients with chronic renal failure are more likely to have a heart condition.

They are particularly vulnerable to infections. Infections are the second leading cause of death in patients with chronic kidney disease [2]. Immunodepression is due to a drop in cellular and humoral immunity. The activity of T and B lymphocytes, monocytes, macrophages and other immune system cells is markedly reduced, given the marked drop in antibody levels [3]. Immune dysfunction is proportional to the duration of chronic renal failure.

As a result, any bacterial infection can rapidly lead to serious septic states, which can result in the patient's death. In fact, catheter-related infections in hemodialysis patients, urinary tract infections in catheter patients and peritonitis in peritoneal dialysis patients are common complications and can cause life-threatening bacteremia.

For decades, antibiotics have helped clinicians to treat infections and improve patients' vital prognosis. Excessive use of these molecules has led to the development of bacterial resistance. In addition, bacterial biofilms, which are bacteria embedded in an extracellular matrix, have the ability to survive in the presence of high concentrations of bactericidal antibiotics and may therefore explain therapeutic failure. Bacterial resistance is a major public health problem, especially as these patients are immunocompromised, have permanent central intravenous catheters or arteriovenous fistulas. This resistance is constantly increasing and is responsible for considerable morbidity and mortality [4]. Bacterial resistance has affected several families of antibiotics, leading to the emergence of the concept of multidrug resistance in 2012. A multi-resistant bacterium (MRB) has been defined as non-sensitivity to at least one agent in three or more classes of antibiotics [5].

Several factors have been incriminated in the acquisition of a BMR, such as renal insufficiency, a central catheter and the length of hospitalisation [6].

Data on the bacterial ecology of the nephrology department is very limited. Few articles on bacterial resistance in patients with kidney disease are available.

The aim of our work is to :

-S To describe the epidemiological profile of the bacterial ecology of the nephrology department: to determine the frequency of bacterial infections in dialysis patients (hemodialysis, peritoneal dialysis), and to identify the germs responsible for bacterial infections,

J To study the resistance profiles of these germs to different antibiotics and to determine

This will enable us to improve the management of nephrology patients by adapting probabilistic antibiotic therapy according to the bacterial ecology found.

1. Type of study

We conducted a retrospective descriptive and analytical study at the Sahloul University Hospital Centre, covering all non-redundant bacterial strains identified from all samples taken from the nephrology department (including the nephrology inpatient unit, the hemodialysis unit and the continuous peritoneal dialysis unit) sent to the microbiology laboratory from January 2012 to December 2020, i.e. a period of 9 years.

2. Collection of bacterial strains :

All non-redundant bacterial strains isolated from samples taken from patients hospitalised in the various units of the nephrology department (nephrology inpatient unit, hemodialysis unit and continuous peritoneal dialysis unit) and sent to the microbiology laboratory at CHU Sahloul.

The samples were divided into groups:
- ECBU
- Blood culture
- Peritoneal fluid
- Equipment: including catheters and urinary catheters
- Collection of superficial pus
- Deep pus sampling
- Ascites fluid
- Coproculture
- Respiratory samples: including sputum and bronchoalveolar fluid (BALF)
- Other samples: including vaginal samples, urethral samples, buccal samples and biopsies.

Only duplicated strains were included in our analyses. In fact, for the same patient, a strain of the same bacterial species with the same antibiotype as a strain already taken into account over a period of less than 15 days is considered to be a duplicate. Samples with a positive yeast culture were also discarded.

3. Bacterial identification and study of

antibiotic resistance :

Bacterial identification was carried out according to :

-S Conventional methods: bacteriological, morphological, cultural, biochemical, enzymatic and antigenic characteristics.

J The VITEK® 2 automated test system (bioMerieux, France).

Antibiotic susceptibility testing was carried out according to the recommendations of the Comité de l'Antibiogramme de la Societe Frangaise de Microbiologie (CA- SFM) from 2012 to 2013, then according to the recommendations of the European Committee on Antimicrobial Susceptibility Testing (CA-SFM/EUCAST) from 2014 to 2020 [7].

In our work, we considered strains with intermediate resistance to be resistant to this antibiotic. We have therefore divided our strains into two categories: either they are sensitive to an antibiotic or they are resistant. A bacterium is considered to be multi-resistant (MDR) if it is resistant to at least three families of antibiotics [5].

Synergy test: looks for the production of extended-spectrum beta-lactamase (ESBL) in enterobacteria.

The ESBL phenotype is detected on the susceptibility test with ceftazidime (30 iig) and cefotaxime (30 iig) discs placed at a distance of 20-30 mm (centre to centre) from an Amoxicillin/Clavulanic Acid (20/10 ng) disc. After incubation for 24 h at 37°C, we will demonstrate a clear increase in the inhibition diameter of discs containing third-generation cephalosporins (C3G) opposite the clavulanic acid/Amoxicillin disc, in the form of a "champagne cork" for ESBL-producing strains.

- E-test strips: for determining minimum inhibitory concentrations (MICs)

- agglutination technique for the detection of PLP2a in meticillin-resistant staphylococci (MRSA)

- use of combined carbapenem+EDTA, carbapenem+boronic acid and carbapenem+clavulanic acid discs to aid identification of carbapenemases

The minimum inhibitory concentrations of colistin for Gram-negative bacilli and vancomycin and teicoplanin for staphylococci were determined using the liquid microdilution method (UMIC biocentric) introduced at the Sahloul University Hospital microbiology laboratory in 2019.

> **Data collection :**

All antibiograms of positive samples from the nephrology hospitalization unit, the hemodialysis (HD) unit and the continuous peritoneal dialysis (CPD) unit over the period from 2012 to 2020 were extracted. This extraction was carried out using the Sahloul microbiology laboratory database. (SIR-Scan and SysLab software).

Duplicates have been eliminated manually.

Antibiotic susceptibility tests for 2015 following an archiving problem in the laboratory.

The data collected for each strain were :

o The requesting department: Nephrology, HD unit, CPD unit

o The date of the sample

o Type of sample: ECBU, blood culture, puncture fluid, deep pus, superficial pus, etc.

o Isolated germ: *Escherichia coli, Klebsiella pneumoniae, Entreococcus faecalis, etc.*

o The level of resistance of each antibiotic tested: sensitive or resistant (intermediate or resistant).

> Statistical analysis of data :

All data were analysed using SPSS version 21 software. The results were presented in the form of graphs, percentages and tables.

The database collected from the SIR-Scan software for the years 2012 to 2017 was opened and organised in Microsoft Excel and then processed by SPSS software.

The data for the period from 2018 to 2020 were transcribed manually into a data collection form using the SysLab software and then transferred to Excel and SPSS.

4. Descriptive study

The statistical unit defined in our study is a bacterial strain.

Qualitative variables were summarised by absolute and relative frequencies. Overall resistance statistics were calculated for the main species of medical interest. For a given bacterial species, the percentage of resistance to an antibiotic was calculated by dividing the number of non-susceptible bacteria by the number of bacteria tested for this antibiotic.

5. Analytical study

The Chi-square test was performed:

- To compare changes in the prevalence of antimicrobial resistance in nephrology between the two periods: before 2016 and after 2016
- The difference is significant if the p-value is less than 0.05 (p<0.05).

6. Ethical considerations

Apart from anonymity, there were no particular ethical considerations for our work.

3 RESULTS

I. Number of isolates per year

A total of 2851 strains were isolated over a 9-year period from 2012 to 2020. The prevalence was stable (0.11) over the last four years of the study. **See Table I**

Table I: Annual incidence of documented infections in the nephrology department

Year	2012	2013	2014	2015	2016	2017	2018	2019	2020
Number of positive samples	321	302	300	366	363	335	360	329	175
Total number of samples						3304	3114	2582	1846
Incidence						0,10	0,12	0,13	0,09

Bacteria were isolated mainly from the nephrology inpatient unit (n=2363, 83%). The numbers of strains collected from the different units are shown in **Figure 1**.

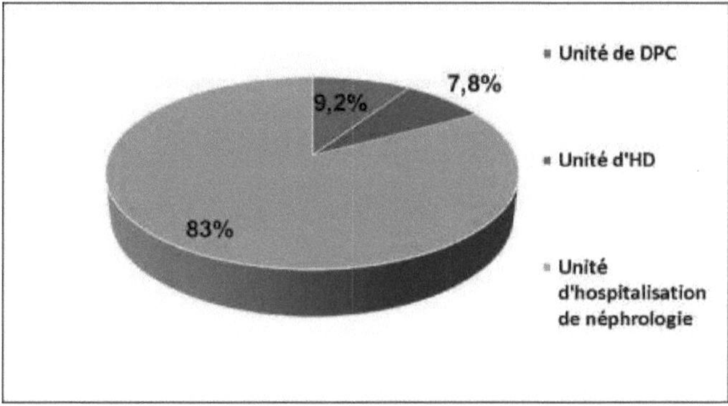

Figure 1: Distribution of isolates by nephrology unit

The number of bacteria identified ranged from 134 to 302 in the nephrology hospital unit. The distribution of the annual number of isolates by hospital unit is **shown in Figure 2.**

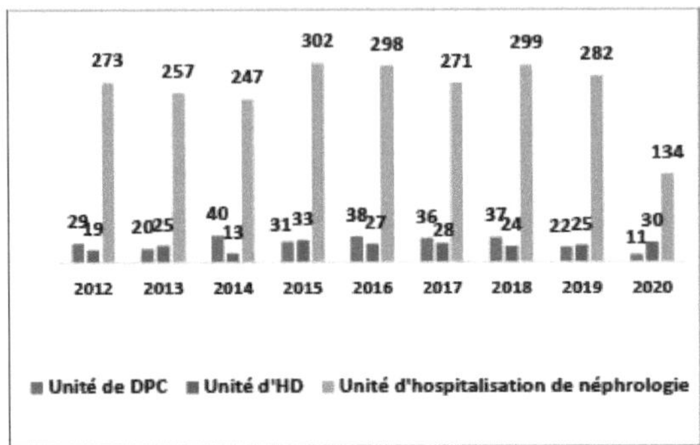

Figure 2: Annual change in the number of isolates in the various units of the nephrology department

II. Distribution of germs per sample

Isolated strains came mainly from cytobacteriological examination of urine (ECBU) (n=1890, **66.3%**), followed by blood cultures (n= 403, **14.1%**), peritoneal fluids (n=181, **6.3%**), material (n=163, 5.7%) and superficial pus samples (n=150, **5.3%**). See Table II

Table II: Distribution of bacteria by sampling site

Collection site	Number	Percentage
ECBU	1890	66,3
Blood culture	403	14,1
Peritoneal fluid	181	6,3
Equipment	163	5,7
Collection of superficial pus	150	5,3
Deep pus sampling	19	0,7
Ascites fluid	17	0,6
Coproculture	15	0,5
Respiratory sampling	8	0,3
Other	5	0,2
Total	2851	100%

: 2 vaginal swabs, 1 urethral swab, 1
Buccal swab and 1 biopsy

111. Distribution of germs by genus and species

Of the 2851 strains, 1893 Gram-negative bacteria (66.4%) and 958 Gram-positive bacteria (33.6%) were isolated.

The majority of germs isolated were enterobacteria (n=1635, 57.3%) followed by Gram-positive cocci (n=950, 33.3%). **See Fig 3**

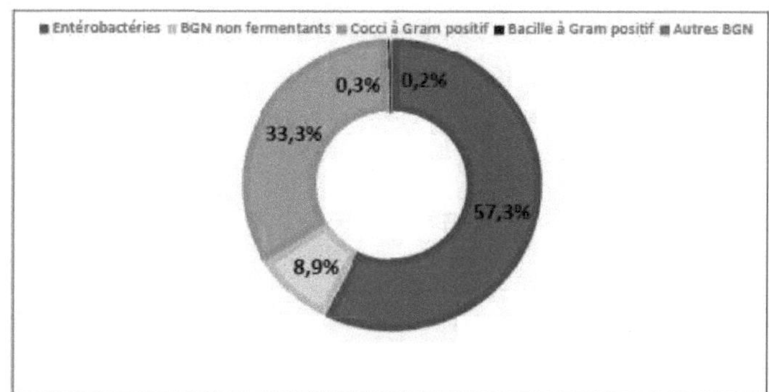

Figure 3: Distribution of isolated bacteria in the nephrology department

In our study, 810 strains of *Escherichia coli* (28.4%), 537 strains of *Klebsiella pneumoniae* (18.8%) and 425 strains of *Staphylococcus aureus* (14.9%) were isolated. (See table III)

Table III: Distribution of isolated bacteria by genus and species

Germ	Number		Percentage
Gram-negative bacilli (GNB)			
Entërobactëries			
Escherichia coli	810		28,4
Klebsiella pneumoniae	537		18,8
Enterobacter cloacae	90	3,2	
Klebsiella oxytoca	35	1,2	
Morganella morganii	31	1,1	
Citrobacter koseri	30	1,1	
Enterobacter aerogenes	20	0,7	
Serratia marcescens	19	0,7	
Proteus mirabilis	16	0,6	
Salmonella sp	13	0,5	
Citrobacter spp	13	0,5	
Proteus sp	05	0,2	
Providencia sp	05	0,2	
Serratia sp	04	0,1	
Pantoea sp	03	0,1	
Raoultella planticola	03	0,1	
Enterobacter sp	01		0,04
Non-fermentative BGN			
Pseudomonas aeruginosa	122	4,3	
Acinetobacter baumannii	54	1,9	
Stenotrophomonas maltophilia	38	1,3	
Pseudomonas sp	15	0,5	
Burkholderia cepacia	08	0,3	
Aeromonas sp	04	0,1	
Acinetobacter sp	02	0,1	

Achromobacter xylosoxidans	02		0,1
Bacteroides sp	02		0,1
Sphingomonas paucimobilis	02		0,1
Elisabethkingia meningoseptica	02		0,1
Coronobacter sakazakii	01		0,04
Other BGN			
Campylobacter sp	02		0,1
Haemophilus parainfluenzae	01		0,04
Actinobacillus sp	01		0,04
Actinobacter sp	01		0,04
Prevotella sp	01		0,04
Gram-positive cocci			
Staphylococci			
Staphylococcus aureus	425		14,9
Coagulase-negative staphylococci			
Staphylococcus epidermidis	45		1,6
Staphylococcus haemolyticus	11		0,4
Staphylococcus hominis	07		0,2
Staphylococcus saprophyticus	02		0,1
Staphylococcus capitis	01		0,04
Staphylococcus caprae	01		0,04
Staphylococcus simulans	01		0,04
Staphylococcus warneri	01		0,04
Streptococci			
Streptococcus agalactiae	55		1,9
***Streptococcus* sp**	23		0,8
Streptococcus pyogenes	14		0,5
***Peptostreptococcus* sp**	01		0,04
Enterococci			
Enterococcus faecalis		253	8,9
Enterococcus faecium		104	3,6
***Enterococcus* sp**	06		0,2
Gram-positive bacilli			
***Corynebacterium* sp**	06		0,2
Brevibacterium casei	01		0,04
Nocardia sp	01		0,04
Total		2851	100%

IV. Distribution of the most frequently isolated germs in function of the sample

J Escherichia coli was the first germ isolated in urine (n=712,37.6%) and the second in positive blood cultures (n=42,10.4%).

J Staphylococcus aureus was the first bacterium isolated from superficial pus samples (n=97, 64%), blood cultures (n=198, 49.1%) and peritoneal fluid cultures (n=43, 23.2%).**see Figure 4.**

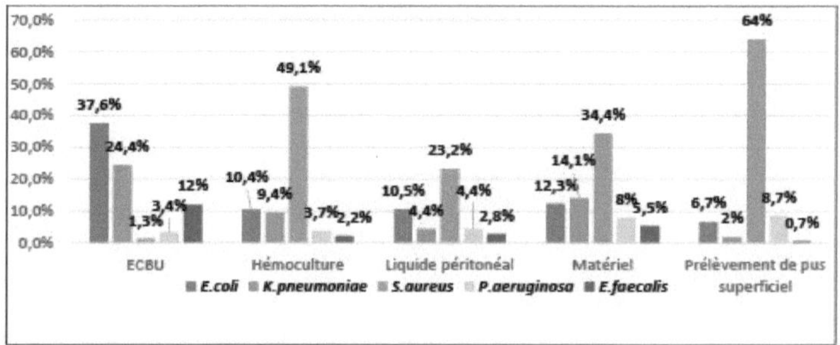

Figure 4: Distribution of the most frequent bacteria according to the sample taken

V. Epidemiology of infections by unit

1. Nephrology inpatient unit

o **Distribution according to sampling site :**

During the study period, 2363 strains were collected at the nephrology inpatient unit. The majority of bacteria isolated in our study came from ECBU (n=1762, 74.6%), followed by blood cultures (n=308, 13%).**See table IV.**

Table IV: Breakdown of infections on the nephrology inpatient unit, by sampling site

Collection site	Number	Percentage
ECBU	1762	**74,6**
Blood culture	308	**13**
Equipment	140	**6,9**
Collection of superficial pus	64	**2,7**
Peritoneal fluid	33	**1,4**
Ascites fluid	16	**0,7**
Coproculture	14	**0,6**
Deep pus sampling	13	**0,6**
Respiratory sampling	08	**0,3**
Other	05	**0,2**
TOTAL	**2363**	**100**

o **Distribution according to the isolated germ:**

In the nephrology department, 71% of the bacteria isolated were gram-negative and 29% gram-positive. Approximately 62.3% of our isolated strains were enterobacteria and 29% were gram-positive cocci. See figure 5

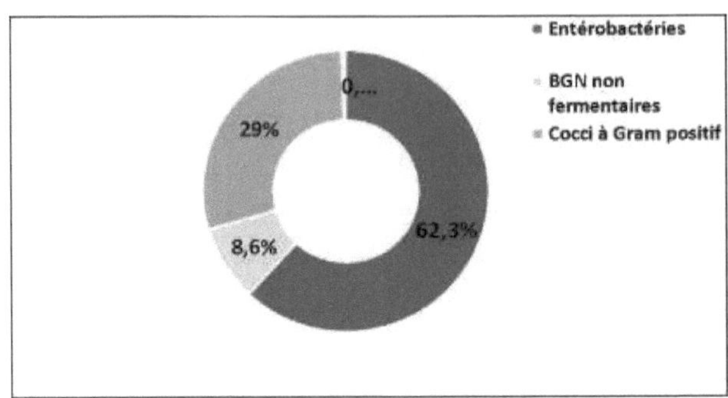

Figure 5: Distribution of isolated bacteria on the nephrology inpatient unit

Escherichia coli (n=734, 31.1%) was the most isolated bacterium on the nephrology inpatient unit, followed by *Klebsiella pneumoniae* (n=501, 21.2%) and *Staphylococcus aureus* (n=244, 10.3%). **See table V**

Table V: Breakdown of bacteria isolated at the nephrology hospital unit by genus and species

Germ	Number		Percentage	
Gram-negative bacilli				
Entbrobactbries				
Escherichia coli	734			31,1
Klebsiella pneumoniae	501			21,2
Enterobacter cloacae		74	3,1	
Morganella morganii		29	1,2	
Klebsiella oxytoca		27	1,1	
Citrobacter koseri		24		1
Enterobacter aerogenes		18	0,8	
Serratia marcescens		15	0,6	
Proteus mirabilis		15	0,6	
Salmonella sp		12	0,5	
Citrobacter sp		12	0,5	
Providencia sp		04	0,2	
Serratia sp		04	0,2	
Pantoea sp		02	0,1	
Raoultella planticola		01	0,04	
Proteus sp		01	0,04	
Non-fermentative BGN				
Pseudomonas aeruginosa		104	4,4	
Acinetobacter baumannii		48	2	
Stenotrophomonas maltophilia		26	1,1	
Burkholderia cepacia		08	0,3	
Pseudomonas sp		06	0,3	
Aeromonas hydrophilia		02	0,1	
Bacteroides sp		01	0,04	

Sphingomonas paucimobilis	01	0,04
Acinetobacter sp	01	0,04
Other BGN		
Campylobacter sp	02	0,1
Haemophilus parainfluenzae	01	0,04
Actinobacillus sp	01	0,04
Actinobacter sp	01	0,04
Prevotella sp	01	0,04
Gram-positive cocci		
Staphylococci		
Staphylococcus aureus	244	10,3
Coagulase-negative staphylococci		
Staphylococcus epidermidis	15	0,6
Staphylococcus haemolyticus	08	0,3
Staphylococcus hominis	05	0,2
Staphylococcus saprophyticus	02	0,1
Staphylococcus caprae	01	0,04
Streptococci		
Streptococcus agalactiae	47	2
Streptococcus sp	11	0,5
Streptococcus pyogenes	06	0,3
Enterococci		
Enterococcus faecalis	238	10,1
Enterococcus faecium	103	4,4
Enterococcus sp	05	0,2
Gram-positive bacilli		
Corynebacterium sp	01	0,04
Nocardia sp	01	0,04
Total	2363	100%

o **Distribution of the most frequently isolated germs according to the collection :**

A total of 1253 enterobacteria (71.1%) were isolated from ECBU at the nephrology hospital. *Escherichia coli* was the first germ isolated from urine (n=656, 37.2%), followed by *Klebsiella pneumoniae* (n=432, 24.5%) and *E.faecalis* (n=223, 12.6%). In the case of blood cultures, Staphylococcus aureus was in the majority (n=140, 45.4%), followed by *Escherichia coli* (n=37, 12%) and *Klebsiella pneumoniae* (n=37, 12%) **(see Figure 6).**

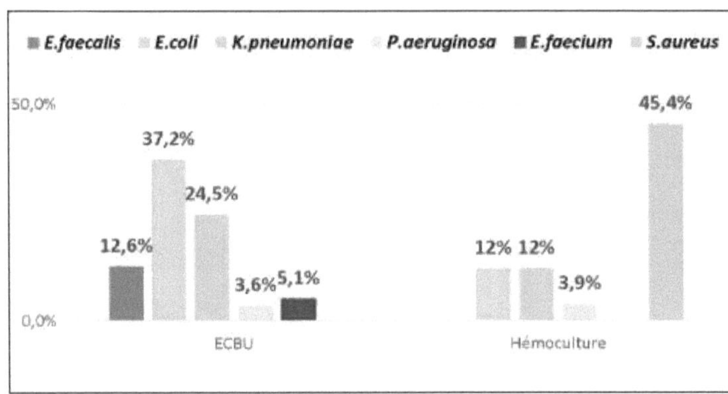

Figure 6: Breakdown of the most frequent bacteria according to the sample taken at the nephrology inpatient unit

2. Dialysis unit (HD)

o **Breakdown by sampling site :**

During the study period, 224 strains were collected at the hemodialysis unit. The majority of strains were isolated from ECBU (n= 84, 37.5%) and blood cultures (n=83, 37.1%).**See table VI.**

Table VI: Breakdown of infections at the HD unit by sampling site

Collection site	Number	Percentage
ECBU	84	37,5
Blood culture	83	37,1
Collection of superficial pus	32	14,3
Equipment	20	8,9
Peritoneal fluid	3	1,3
Deep pus sampling	2	0,9
TOTAL	224	100

o **Distribution according to isolated germ:**

At the hemodialysis unit, 56.7% of the bacteria isolated were gram-positive and 43.3% were gram-negative. Approximately 55.8% of the bacteria isolated were gram-positive cocci. Enterobacteriaceae were in second place (n=88, 39.3%).**see figure 6.**

14

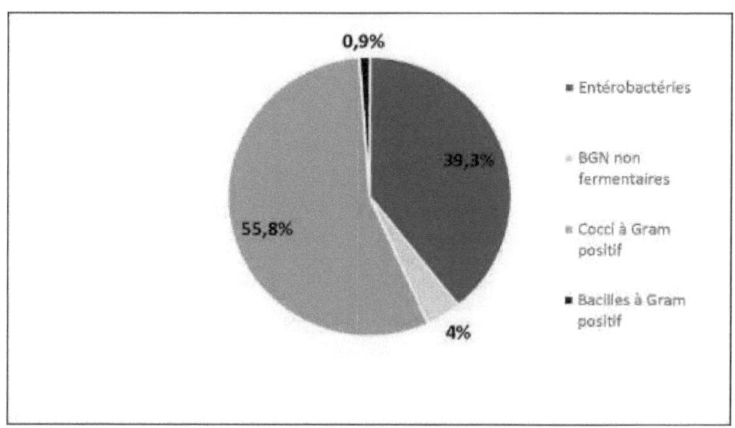

Figure 7: Distribution of bacteria isolated at the HD unit

Staphylococcus aureus (n=87, 38.8%) was the most prevalent bacterium in the HD unit, followed by *Escherichia coli* (n=39, 17.4%). **See table VII**

Table VII: Distribution of bacteria isolated at the HD unit by genus and species

Germ	Number	Percentage
Gram-positive cocci		
Staphylococci		
Staphylococcus aureus	87	38,8
Coagulase-negative staphylococci		
Staphylococcus epidermidis	14	6,3
Staphylococcus hominis	01	0,4
Staphylococcus simulans	01	0,4
Streptococci		
Streptococcus agalactiae	05	2,2
Streptococcus sp	02	0,9
Streptococcus pyogenes	07	3,1
Enterococci	08	3,6
Enterococcus faecalis		
Gram-negative bacilli		
Enterobacteria		
Escherichia coli	39	17,4
Klebsiella pneumoniae	21	9,4
Enterobacter cloacae	07	3,1
Klebsiella oxytoca	06	2,7
Morganella morganii	02	0,9
Citrobacter koseri	02	0,9
Enterobacter aerogenes	02	0,9
Serratia marcescens	02	0,9
Proteus mirabilis	01	0,4
Citrobacter sp	010 ,4	
Proteus sp	031 ,3	

Providencia sp	010 ,4
Enterobacter sp	010 ,4
Non-fermentative BGN	041 ,8
Pseudomonas aeruginosa	010 ,4
Acinetobacter baumannii	010 ,4
Aeromonas sp	020,9
Achromobacter xylosoxidans	010 ,4
Bacteroides sp	
Gram-positive bacilli	
Corynebacterium sp	020,9
Total	224100%

Analysis of the ECBU of hemodialysis patients during the period of our study showed a predominance of enterobacteria, essentially *Escherichia coli* (n=34, 40.5%) followed by *Klebsiella pneumoniae* (n=19, 22.6%). The bacteriological profile of the blood cultures showed a predominance of *Staphylococcus aureus* (n=51, 61.5%) followed by *Staphylococcus epidermidis* (n=10, 12%). **See figure 8**

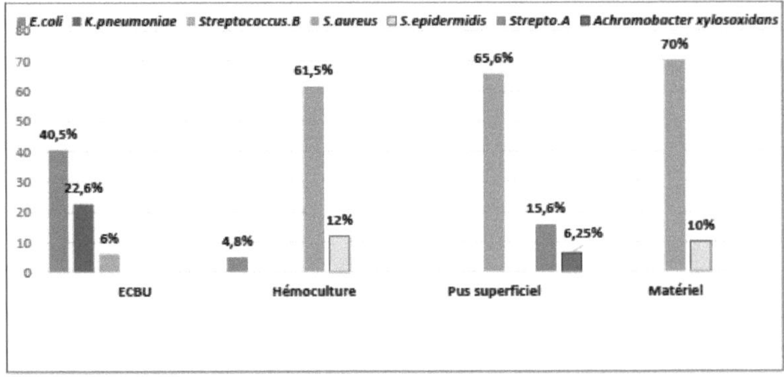

Figure 8: Distribution of the most frequently isolated bacteria as a function of the sampling at the HD unit

3. Continuous peritoneal dialysis unit (CPD)

○ **Breakdown by sampling site :**

During the period of our study, 264 isolates were collected. The majority of our strains were isolated from peritoneal fluid samples (n=145, 54.9%) followed by superficial pus samples (n=54, 20.5%) and other samples detailed in **Table VIII**.

Table **VIII**: Breakdown of infections at the DPC unit by sampling site

Collection site	Number	Percentage
Peritoneal fluid	145	54,9
Collection of superficial pus	54	20,5

ECBU	44	16,7
Blood culture	12	4,5
Deep pus sampling	4	1,5
Equipment	3	1,1
Ascites fluid	1	0,4
Coproculture	1	0,4
TOTAL	264	100

o **Distribution according to the isolated germ:**

At the peritoneal dialysis unit, 55% of the bacteria isolated were Gram-positive and 45% were Gram-negative.

The majority of isolates were gram-positive cocci (53%), followed by enterobacteria (28%). **See figure 9**

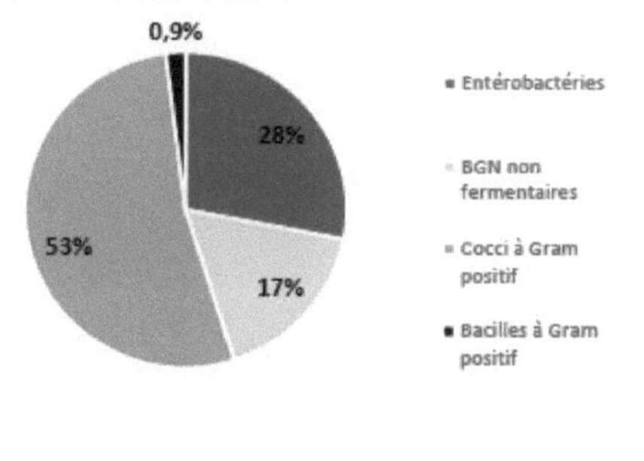

Figure 9: Distribution of bacteria isolated at the DPC unit

In the peritoneal dialysis unit, *Staphylococcus aureus* (n=97, 35.6%) was the most isolated bacterium, followed by *Escherichia coli* (n=37, 14%) and *Staphylococcus epidermidis* (n=16, 6.1%). **See table IX**

Table IX: Distribution of bacteria isolated at the DPC unit by genus and species

Germ	Number	Percentage
Gram-positive cocci		
Staphylococci		
Staphylococcus aureus	94	35,6
Coagulase-negative staphylococci		
Staphylococcus epidermidis	16	6,1
Staphylococcus haemolyticus	03	1,1
Staphylococcus hominis	01	0,4
Staphylococcus capitis	01	0,4
Staphylococcus warneri	01	0,4
Streptococci		
Streptococcus agalactiae	03	1,1
Streptococcus sp	10	3,8
Streptococcus pyogenes	01	0,4
Peptostreptococcus sp	01	0,4

Enterococci		
Enterococcus faecalis	07	2,7
Enterococcus faecium	01	0,4
Enterococcus sp	01	0,4
Gram-negative bacilli		
Enterobacteria		
Escherichia coli	37	14
Klebsiella pneumoniae	15	5,7
Enterobacter cloacae	09	3,4
Citrobacter koseri	04	1,5
Klebsiella oxytoca	02	0,8
Serratia marcescens	02	0,8
Raoultella planticola	02	0,8
Salmonella sp	01	0,4
Proteus sp	01	0,4
Pantoea sp	01	0,4
Non-fermentative BGN		
Pseudomonas aeruginosa	14	5,3
Stenotrophomonas maltophilia	12	4,5
Pseudomonas sp	09	3,4
Acinetobacter baumannii	05	1,9
Elisabethkingia meningoseptica	02	0,8
Aeromonas sp	01	0,4
Sphingomonas paucimobilis	01	0,4
Coronobacter sakazakii	01	0,4
Gram-positive bacilli		
Corynebacterium sp	03	1,1
Brevibacterium casei	02	0,8
TOTAL	264	100%

At DPC, *Staphylococcus aureus* was the most isolated bacterium in peritoneal fluid, superficial pus and blood cultures. *Escherichia coli* was the most common bacterium in positive cultures from ECBU. **See table X**

Table X: Breakdown of the most frequently isolated bacteria according to the sampling at the DPC unit

Collection site	Isolated bacteria
Peritoneal fluid	*S.aureus* 23.4% *E.coli* 10.3 *S.epidermidis* 9%
More superficial	*S.aureus* 85.2% *P.aeruginosa* 11.1%
ECBU	*E.coli* 47.7 *K.pneumoniae* 22.7%
Blood culture	*S.aureus* 58.3% *E.coli* 8.3% *S.epidermidis* 8.3%

Bacteria isolated from the peritoneal fluid of patients admitted to the DPC unit are **shown in Figure 10.**

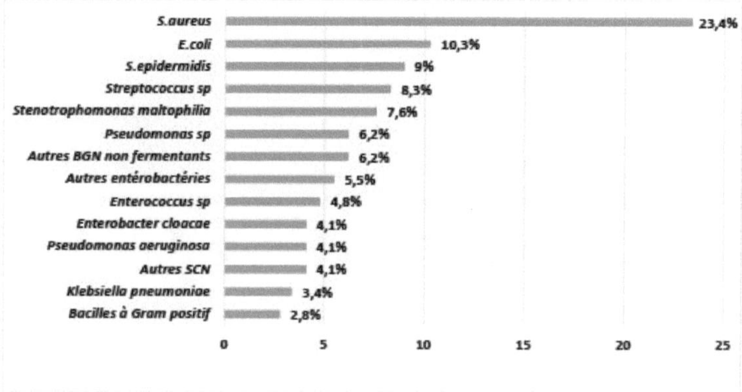

Figure 10: Distribution of bacteria isolated from peritoneal fluid at the DPC unit

VI. Antibiotic resistance in the main bacteria

1. Enterobacteria

o *Escherichia coli :*

Escherichia coli resistance to amoxicillin was approximately 82.2%, and to the combination of amoxicillin and clavulanic acid 54.6%. For fluoroquinolones, resistance to ciprofloxacin was 47.8%. All *E.coli* strains were sensitive to colistin. **See table XI.**

Tableau XI: Study of *Escherichia coli* resistance to antibiotics

	AMX (n/N')	AMC (n/N')	CTX (n/N')	IMP (n/N')	FTE (n/N')	GN (n/N')	AMK (n/N')	CIP (n/N')	SXT (n/N')	Fos (n/N')	CT (n/N')
Total N=724	595/724 (82,2%)	395/724 (54,6%)	248/723 (34,3%)	2/715 (0,3%)	14/714 (2%)	155/713 (21,7%)	63/722 (8,7%)	346/724 (47,8%)	383/699 (54,8%)	13/673 (1,9%)	0/356 (0%)

AMX: Amoxicillin; AMC: Amoxicillin-Clavulanic Acid; CTX: Cefotaxime; IMP: Imipeneme ;
AMK: Amikacin; CIP: Ciprofloxacin; SXT: Cotrimoxazole; Fos: Fosfomycin; CT: Colistin

o *Klebsiella pneumoniae :*

Resistance of *Klebsiella pneumoniae* to Amoxicillin-Clavulanic Acid and ciprofloxacin was 61% and 59.5% respectively. Approximately 2.4% of strains were resistant to colistin. **See table XII**

Tableau XII: Study of the resistance of *K.pneumoniae* to antibiotics

	AMC (n/N')	CTX (n/N')	IMP (n/N')	FTE (n/N')	GN (n/N')	AMK (n/N')	CIP (n/N')	SXT (n/N')	Fos (n/N')	CT (n/N')
Total N=724	277/454 (61%)	274/452 (60,6%)	54/449 (12%)	92/448 (20,5%)	175/450 (38,9%)	54/452 (11,9%)	256/446 (59,4%)	230/433 (53,1%)	26/446 (5,8%)	6/254 (2,4%)

AMC: Amoxicillin-Clavulanic Acid; CTX: Cefotaxime; IMP: Imipeneme; ETP: Ertapeneme; GN: Gentamicin; AMK: Amikacin; CIP: Ciprofloxacin; SXT: Cotrimoxazole; Fos: Fosfomycin; CT:

Colistin

2. Staphylococci

o *Staphylococcus aureus* :

Staphylococcus aureus resistance to penicillin G was 89.7%. Resistance to erythromycin and lincomycin was 11.8% and 1.4% respectively. **See table XIII**

Table XIII: Study of the resistance of S.aureus to antibiotics

	PG (n/N')	OXA (n/N')	KAN (n/N')	GN (n/N')	ERY (n/N')	LNC (n/N')	PTN (n/N')	OFX (n/N')	TEC (n/N')	SXT (n/N')	AF (n/N')
Total N=369	331/369 (89,7%)	48/369 (13%)	49/368 (13,3%)	10/367 (2,7%)	42/355 (11,8%)	5/363 (1,4%)	0/355 (0%)	12/360 (53,1%)	4/324 (1,2%)	5/363 (1,3%)	71/369 (19,2%)

PG: Penicillin G; OXA: Oxacillin; KAN: Kanamycin; GN: Gentamicin; ERY: Erythromycin; LNC: Lincomycin;
PTN: Pristinamycin ; OFX: Ofloxacin ; TEC: Teicoplanin ; SXT: Cotrimoxazole ; AF : Fusidic acid

3. *Pseudomonas* sp

o *Pseudomonas aeruginosa* :

Resistance of *Pseudomonas aeruginosa* to ticarcillin and to the combination of ticarcillin and clavulanic acid was 45.6% and 36.8% respectively. For fluoroquinolones, resistance to ciprofloxacin was 27.7%. All isolated strains of *P.aeruginosa* were sensitive to colistin. **See table XIV**

Table XIV: Study of *P. aeruginosa* resistance to antibiotics

	ICT (n/N')	CBT (n/N')	PIP (n/N')	PIPTZ (n/N')	CAZ (n/N')	IMP (n/N')	CIP (n/N')	AMK (n/N')	Fos (n/N')	CT (n/N')
Total N=101	42/92 (45,6%)	32/87 (36,8%)	24/98 (24,5%)	15/98 (15,3%)	14/97 (14,4%)	8/100 (8%)	28/101 (27,7%)	8/101 (7,9%)	6/51 (11,8%)	0/53 (0%)

TIC: Ticarcillin; TCC: Ticarcillin-Clavulanic Acid; PIP: Piperacillin; PIPTZ: Piperacillin-Tazobactam; CAZ: Ceftazidime; IMP: Imipenem; CIP: Ciprofloxacin; AMK: Amikacin; Fos: Fosfomycin; CT: Colistin

4. *Acinetobacter baumannii*

Resistance of *A.baumannii* to ticarcillin and to the combination of ticarcillin and clavulanic acid was 77.5% and 80.5% respectively. Resistance to piperacillin and tazobactam was 72.4%. **See table XV**

Table XV: Study of the resistance of *A.baumannii* to antibiotics

	ICT (n/N')	CBT (n/N')	PIP (n/N')	PIPTZ (n/N')	CAZ (n/N')	IMP (n/N')	CIP (n/N')	AMK (n/N')	SXT (n/N')	CT (n/N')
Total N= 38	31/40 (77,5%)	29/36 (80,5%)	26/32 (81,2%)	21/29 (72,4%)	27/34 (79,4%)	16/40 (40%)	33/40 (82,5%)	21/40 (52,5%)	19/38 (50%)	0/30 (0%)

TIC: Ticarcillin; TCC: Ticarcillin-Clavulanic Acid; PIP: Piperacillin; PIPTZ: Piperacillin-Tazobactam; CAZ: Ceftazidime; IMP: Imipenem; CIP: Ciprofloxacin; AMK: Amikacin; CT: Colistin

5. Enterococci

o *Enterococcus faecalis* :

Resistance of *Enterococcus faecalis* to ampicillin was 2%. Resistance to erythromycin was 88.1%. Approximately 2% of strains were resistant to glycopeptides. **See table XVI**

Tableau XVI: Study of global antibiotic resistance in *E. faecalis*

	AMP (n/N')	ERY (n/N')	LEV (n/N')	GN (n/N')	KAN (n/N')	VAN (n/N')	TEC (n/N')	SXT (n/N')	RIF (n/N')	FOS (n/N')
Total N= 210	3/152 (2%)	185/210 (88,1%)	19/190 (55,9%)	117/153 (76,5%)	128/153 (83,7%)	2/204 (2%)	4/210 (1,9%)	122/201 (60,7%)	31/157 (19,7%)	93/162 (57,4%)

AMP: Ampicillin; ERY: Erythromycin; LEV: Levofloxacin; GN: Gentamycin; VAN: Vancomycin; TEC: Teicoplanin; SXT: Cotrimoxazole; RIF: Rifamycin;FOS: Fosfomycin

o *Enterococcus faecium* :

Resistance of *Enterococcus faecium* to erythromycin and lincomycin was 96.2% and 88.5% respectively. Glycopeptide resistance was 25.4%. **See table XVII**

Tableau XVII: Study of global antibiotic resistance in *E. faecium*

	ERY (n/N')	LNC (n/N')	PTN (n/N')	LEV (n/N')	GN (n/N')	KAN (n/N')	VAN (n/N')	TEC (n/N')	SXT (n/N')	RIF (n/N')
Total N= 80	76/79 (96,2%)	46/52 (88,5%)	4/65 (6,2%)	24/60 (40%)	44/47 (93,6%)	61/61 (100%)	12/69 (17,4%)	20/79 (25,4%)	58/61 (95,1%)	37/47 (78,7%)

ERY: Erythromycin ; LNC: Lincomycin ; PTN: Pristinamycin ; LEV: Levofloxacin; GN: Gentamycin; KAN: Kanamycin; VAN: Vancomycin; TEC: Teicoplanin; SXT: Cotrimoxazole; RIF: Rifamycin

VII. Prevalence of multi-resistant bacteria in the department of nephrology

Among the 2851 strains isolated in nephrology, 688 multi-resistant bacteria were found, representing 24.1% of isolates. The rate of multi-resistant BGN in our study was 21.5% (n=614). **See Table XIX**

Tableau XVIII: Prevalence of BMR

BMR	N	N'	Prevalence
EntRC3G	457	1464	31,2%
ERC	126	1454	8.7%
ERG	25	304	8,2%
SHELTER	16	40	40%
PseudoCAZR	15	112	13,4%
MRSA	49	380	12,9%

N : Number of multi-resistant strains; N': Total number of strains

VIII. Distribution of isolated BMR in the nephrology department

In the nephrology department, 457 isolated strains of enterobacteria (31.2%) were resistant to C3Gs and only 126 strains were resistant to carbapenems (8.7%). For *S.aureus*, 49 strains were resistant to meticillin (12.9%). For *Acinetobacter baumannii*, 16 strains were resistant to imipenem (40%). For *Pseudomonas* sp, resistance to ceftazidime was 15.2% (n=15). Enterococcal resistance to glycopeptides was 8.2% (n=25). The distribution of these BMR to the different units of the department is detailed **in figure 11.**

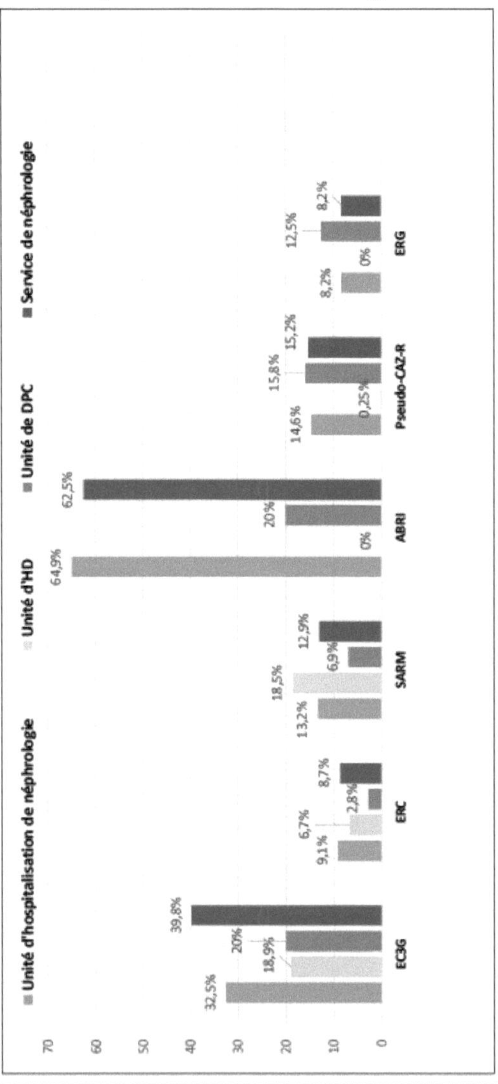

EC3G: Enteribacteria resistant to third generation cephalosporins ; ERC: enterobacteria resistant to caϭbapënëmex ; MRSA: staphylococcus aureus lAsistant to mëbcИИne ; ABRI: Acinetobacter baumannii lAsistant to Гmupënëme ; Pseudo-CAZ-R: Pseudomonas sp lAsistant to cëftazidime ; ERG: Enterococci lAsistant to glycopeptides

Figure 11: Distribution of BMR in the different units of the nephrology department

22

VII. Trend in the number of BMR in nephrology by year

The evolution of resistance of enterobacteria to C3G during the period of our study was stable. Resistance to carbapenemes in these bacteria halved between 2016 and 2017. **See figure 12**

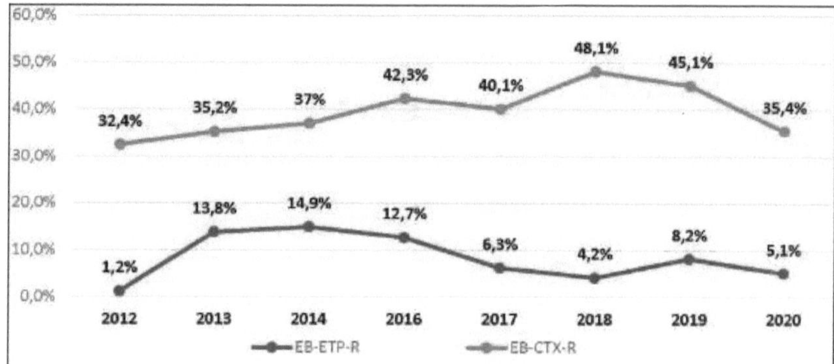

EB-CTX-R: Enterobacteria resistant to cefotaxime; EB-ETP-R: Enterobacteria resistant to ertapenem.

Figure 12: Overall annual trend in enterobacteria resistant to C3G and carbapenems

With regard to *Pseudomonas* sp, an increase in the rate of resistance to ceftazidime was noted after 2016. The number of ABRIs doubled between 2016 and 2018.

See figure 13

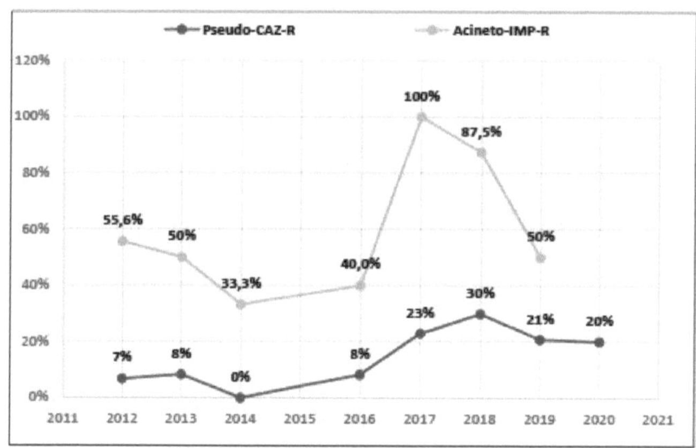

Pseudo-CAZ-R: Ceftazidime-resistant Pseudomonas spp; Acineto-IMP-R: Imipenem-resistant Acinetobacter baumannii

Figure 13:Annual trends in resistance of *Pseudomonas* sp to ceftazidime and Acinetobacter baumannii to imipenem.
***Acinetobacter baumannii* to imipeneme**

Glycopeptide-resistant enterococci appeared in 2016. The
MRSA prevalence has tripled from 2016 to 2017. **See figure 14**

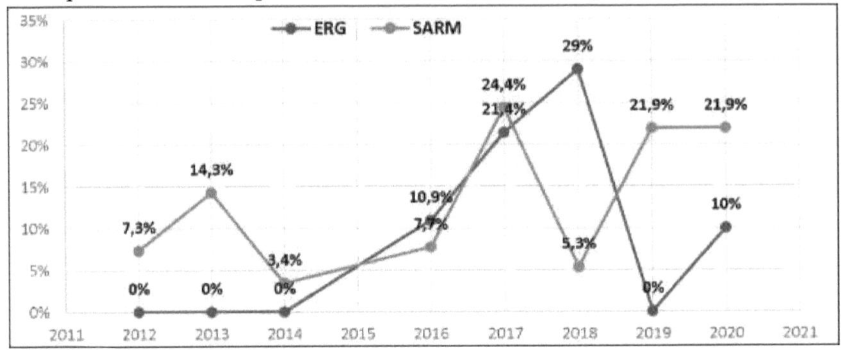

MRSA: meticillin-resistant staphylococcus aureus; GREE: glycopeptide-resistant Enterococci.

Figure 14: Annual trends in resistance to meticillin *in S. aureus* and to glycopeptides in enterococci to glycopeptides

However, the difference in prevalence between the two periods (before and after 2016) was statistically significant for the following MRBs: C3G-resistant enterobacteria, glycopeptide-resistant enterococci, MRSA and ceftazidime-resistant *P.aeruginosa* **(Table XIX)**.

Table XIX: Trends in the prevalence of antimicrobial resistance in nephrology

		Period		Total	P
		Before 2016	After 2016	(100%)	
Enterobacteria and C3G	Rësistant	146 (31,9%)	311 (68,1%)	457	<0.001
	Sensitive	1143 (47,7%)	1251 (52,3%)	2394	
Enterobacteria and Carbapenemes	Rësistant	59 (46,8%)	67 (53,2%)	126	0.7
	Sensitive	1230 (45,1%)	1495 (54,9%)	2725	
Enterococci and glycopeptides	Rësistant	25	0	25 (100%)	<0.001
	Sensitive	1289 (45,6%)	1537 (54,4%)	2826	
S. aureusand meticillin	Rësistant	9 (18,4%)	40 (81,6%)	49	<0.001
	Sensitive	1280 (45,7%)	1522 (54,3%)	2802	
Acinetobacter baumanniiand	Rësistant	(43,8%)	(56,3%)	16	0.9

imipeneme	Sensitive	1282 (45,2%)	1553 (54,8%)	2835	
Pseudomonas sp and ceftazidime	Rësistant	213 (13,3%)	(86,7%)	15	0.013
	Sensitive	1287 (45,4%)	1549 (54,6%)	2836	

4 DISCUSSION

Nephrology patients are fragile individuals with a number of risk factors for developing a bacterial infection (immunodepression, kidney transplants, haemodialysis, peritoneal dialysis). In fact, this risk is a hundred times higher than in the general population. This complication is the second leading cause of death in these patients [8]. The aim of our work is to identify the germs involved in bacterial infections in the nephrology department, to study the resistance profiles of these germs to the various antibiotics and to determine the frequency of bacterial infections in dialysis patients (hemodialysis, peritoneal dialysis) in order to improve the management of nephrology patients by adapting the probabilistic antibiotic therapy according to the bacterial ecology of the department.

In our study, 2851 strains were isolated over a period from January 2012 to December 2020. The majority of germs were collected from the nephrology inpatient unit (n=2363, 83%), followed by the CPP unit (n=264, 9.2%) and the HD unit (n=224, 7.8%). Most isolates were identified from ECBU (n=1890, 66.3%) followed by blood cultures (n=403, 14.1%). At the nephrology hospital unit, n=1676 BGN were isolated out of a total of 2363, a rate of 71%. Among these BGN, enterobacteria were in the majority (n=1473, 61.3%). *Escherichia coli* (n=734, 31.1%) ranked first, followed by *Klebsiella pneumoniae* (n=501, 21.2%). In both the HD and DPC units, Gram-positive cocci predominated (n=125 or 55.8% in HD and n=140 or 53% in DPC). *Staphylococcus aureus* was the main germ found in the HD unit (n=87, 38.9%) and in the DPC unit (n=94, 35.6%). Among the 2851 strains isolated in nephrology, 688 multi-resistant bacteria were found, representing 24.13% of isolates. C3G-resistant enterobacteria were the most common type of MRB (n=457, 16%), followed by carbapenem-resistant enterobacteria (n=126, 4.4%). In third place was MRSA (n=49, 1.73%), followed by glycopeptide-resistant enterococci (n=25, 0.9%), imipenem-resistant *A.baumannii* (n=16, 0.6%) and ceftazidime-resistant *Pseudomonas* sp. (n=15, 0.5%).

I. Number of isolates per year :

The average prevalence was stable at around 0.11. The average number of positive samples taken in nephrology was around 317 per year. This number fell in 2020 to just 175 positive samples. However, the prevalence (which is the number of positive samples out of the total number of samples taken in 2020) is 0.09. This could be explained by the drop in the number of positive samples. This could be explained by the drop in samples taken during the first wave of the COVID-19 pandemic. A study carried out in the Nouvelle Aquitaine region showed a 50% drop in the use of emergency care during the first wave of containment and a 30% drop during the second wave of containment [9]. A study by Heist et al. in 2021 noted a drop in the total number of hospital admissions to 69.2% of planned admissions in the week ending 4 April 2020 [10].

II. Breakdown of infections on the nephrology inpatient unit, by sample and by germ

Urinary tract infections ranked first (n=1762, 74.6%) followed by bacteremia (n=308, 13%). Bacteremia in these patients may have a urinary origin. The predominance of urinary strains could be due to the antecedents of nephrology patients who generally have underlying urological pathologies such as chronic renal failure or the insertion of an indwelling urinary catheter. Our results are similar to those of a study carried out in the Nephrology and Renal Transplantation Departments of the Imam Khomeini Hospital Complex, Iran [11], which showed a predominance of positive cultures from ECBU (79 samples, 72.5%) followed by blood cultures (17 samples, 15.6%).

1. Distribution of uropathogenic germs :

The majority of urinary bacteria isolated in nephrology were enterobacteria (n=1253, 71.1%). *Escherichia coli* was found to be the most prevalent bacterium (n=656, 37.2%) followed by *Klebsiella pneumoniae* (n=432, 24.5%). This is in line with the results of Samanipour et al. in Iran [11].

From a global point of view, *E. coli* is the first germ to be isolated in the microbiology laboratories of various Tunisian [12] or international [13] hospitals. *Escherichia coli*, a commensal of the digestive tract, is responsible for ascending urinary tract infections. Certain pathovars acquire virulence factors such as adhesins and *pili*, which reinforce colonisation of the urinary epithelium [14]. Our results are consistent with most of the studies reported in **Table XX**.

Table XX: Comparison of the distribution of germs involved in urinary tract infections infections in certain hospitals

Authors	Date and place	Result
Wang et al [15]	China (2013-2015)	*E.coli* 30.3 *K.pneumoniae* 22
Shrimali et al [16]	Gandhinagar, India (2018)	*E.coli* 34.8 *P.vulgaris* 15.9% *S.saprophyticus* 14
Adhikary et al [17]	Kathmandu, Nepal (2017)	*E.coli* 64 *Proteus* sp 12.5 *Enterobacter* sp 12.5% *Klebsiella* sp 11
Tayh et al [18]	Gaza, Palestine (2013)	*E.coli* 70.6 *K.pneumoniae* 17.6%

		P.mirabilis 3.5
		E.colacae 3.5
Mawufemo et al [19]	Lome, Togo (2017-2018)	*E.coli* 59.2%
		Klebsiella sp 27.8
Chemlal et al [20]	Oujda, Morocco (2012-2014)	*E.coli* 54.5
		Klebsiella sp 30.9
		StreptoD 5.5
Guermazi-Toumi et al [21].	Gafsa, Tunisia (2015-2016)	*E.coli* 64
		K.pneumoniae 12.8%
		P.mirabilis 2.6%
Our study	**Sousse, Tunisia (2012-2020)**	*E.coli* 37.2%
		K.pneumoniae 24.5

2. Distribution of germs isolated in blood cultures

In our work, the most isolated germs in blood cultures were *S.aureus* (n=140, 45.4%) followed by *E.coli* (n=37,12%) and *K*.pneumoniae (n=37, 12%).

Staphylococcal haemocultures could enter through a mucocutaneous infection or a catheter infection in hemodialysis patients. The patients in our study are immunocompromised due to chronic renal failure, and these immunocompromised patients could present bacteremia due to the imbalance of their own opportunistic pathogenic commensal cutaneous-mucosal flora [22]. In contrast, enterobacteria-positive blood cultures may follow urinary tract infections in patients with bladder catheters [23].

Our results are consistent with the multicentre study conducted by Rosenthal et al. between 2013 and 2019 [24]. In studies carried out in Ireland and South Asia [25,26], staphylococci are monitored by isolating strains *of P.aeruginosa* and Beta-hemolytic streptococci from blood cultures.

The germs isolated from blood cultures in the various studies are shown in **table XXI.**

Table XXI: Comparison of the distribution of germs involved in bacteremia in certain hospitals

Authors	Date and place	Result
Zhang et al [27]	Texas, United States (2012-2015)	*S.aureus* 41.6%
		Enterobacter spp 16.8%
		SCN 13.9%
Fram et al [28]	Sao Paulo, Brazil (2010-2013)	*S.aureus* 32.1%
		S.epidermidis 13.6%
		Enterobacteria 25.9%
		SCN 30%
Agrawal et al [26]	South Asia (2016-2017*)*	*S.aureus 13.2*
		P.aeruginosa 11.3%
Mohamed et al [25]	Ireland (2015-2016)	SCN 61.7

		S.aureus 23.4%
		Beta-hemolytic S. 4.3%
Alhazmi et al [29]	Jeddah, Saudi Arabia (2014-2016)	SCN 18.2%
		K.pneumoniae 15.2%
		S.aureus 9.1%
Rosenthal et al [24]	Bahrain, Egypt, Iran, Jordan, Saudi Arabia, Kuwait, Lebanon, Morocco, Pakistan, Palestine, Sudan, Tunisia, Turkey and United Arab Emirates (2013-2019)	SNA 31 *S.aureus* 14% *K.pneumoniae* 8% *E.coli* 7%
Our study	**Sousse, Tunisia (2012-2020)**	**S.aureus 49.1%** **E.coli 10.4** **K.p neumoniae 9.4** SCN 6%

III.Distribution of infections per HD unit per sample and by germ

Analysis of the ECBU of hemodialysis patients during the period of our study showed a predominance of enterobacteria, essentially *Escherichia coli* (n=34, 40.5%) followed by *Klebsiella pneumoniae* (n=19, 22.6%).

This result is consistent with most of the epidemiological studies found, which are shown in **table XXII**.

Table XXII: Comparison of the distribution of germs involved in urinary tract infections
in some hemodialysis units

Authors	Location and date	Result
Yamashita et al [30]	Tokyo, Japan (2020-2021)	*Escherichia coli 62.5* *Klebsiella pneumoniae 12.5* *Enterococcus faecium 12.5* *Streptococcus group B 10% of the total*
Menhal et al [31]	Baghdad, Iraq (2010-2011)	*Escherichia coli 40% of the total* *Klebsiella* spp 33.3%
Rahman et al [32]	Dhaka, Bangladesh (2021)	*Escherichia coli 51.8* *Staphylococcus aureus 28.8* *Staphylococcus saprophyticus 11.2* *Klebsiella* spp 8.2% *Enterococcus* spp 4.7* ^{O}
Our study	Sousse, Tunisia (2012-2020)	*Escherichia coli 40.5* *Klebsiella pneumoniae 22.6* *Streptococcus group B 6%*

In hemodialysis patients, a study of the bacteriological profile of blood cultures showed a predominance of Gram-positive cocci, with Staphylococcus aureus (n=51, 61.5%) in first place, followed by *Staphylococcus epidermidis* (n=10, 12%). The predominance of *Staphylococcus aureus* could be explained by nasal carriage of this bacterium, which is then transmitted to the catheter insertion site by contaminated hands, leading to bacteremia. Prophylaxis of nasal carriage with mupirocin is therefore desirable in hemodialysis patients in order to reduce the risk of developing an *S.aureus* catheter infection complicated by sepsis [33].

Table XXIII: Comparison of the distribution of germs involved in bacteremia in bacteremia in certain hemodialysis hospitals

Authors	Date and place	Result
Bhojaraja et al [34]	Karnataka, India (20172018)	*SNA 24.6* *Staphylococcus aureus 18* *Klebsiella pneumoniae 11.5* *Enterococcus* spp 9.8% *Escherichia coli 8.2%*
Alhazmi et al [29]	Jeddah, Saudi Arabia (2014-2016)	*SCN 18.2%* *Klebsiella pneumoniae 15.2%* *Staphylococcus aureus 9.1*
Abd El-Hamid El-Kady et al [35]	Jeddah, Saudi Arabia (2019-2020)	*Staphylococcus aureus 30.2* *Staphylococcus epidermidis 27.2* *Klebsiella pneumoniae 10.8* *Pseudomonas aeruginosa 8.6*
Krishnan et al [36]	Perth, Australia (2010 2014)	*Staphylococcus aureus 31.4* *SCN 17.5* *Pseudomonas aeruginosa 8.7*
Our study	**Sousse, Tunisia (2012) 2020)**	*Staphylococcus aureus 61.5* *Staphylococcus epidermidis 12* *Escherichia coli 4.8*

IV.Microbiological profile of infections at the DPC unit

Half of the positive cultures at the CPD unit come from peritoneal fluid. Indeed, peritonitis is the first complication encountered in peritoneal dialysis patients.

In this unit, bacterial ecology is dominated by Gram-positive bacteria (n=145, 55%), mainly *S.aureus* (n=57, 21.7%). This germ forms a biofilm that colonises peritoneal catheters and may infect the peritoneal fluid [37,38]. *Staphylococcus epidermidis* followed by other staphylococci are the most frequently isolated bacteria in peritonitis in peritoneal dialysis patients worldwide. The bacteriological profile may vary according to geography. Gram-negative bacteria are more frequent in India and their prognosis is worse than that of Gram-positive peritonitis [39,40].**See table XXIV**.

Table XXIV: Comparison of the distribution of germs involved in peritonitis in

peritoneal dialysis
patients

Authors	Date and place	Result
Ghali et al [41]	Australia (2003-2008)	*SNA 27.2%* *Staphylococcus aureus 12.9* *Streptococcus sp 6.9* *Escherichia coli 6.3%*
Wang et al [42]	Taiwan (2007-2016)	*SCN 17.4%* *Streptococcus sp 15.8* *Staphylococcus aureus 7.9* *Escherichia coli 6.3%*
Song et al [43]	Changsha, China (2014-2020)	*Staphylococcus epidermidis 11.1* *Streptococcus sp 10.5* *Escherichia coli 10.1* *Staphylococcus aureus 9.8*
Kanjanabuch [44]	Bangkok, Thailand (2016-2017)	*Streptococcus sp 16* *SCN 12%* *Staphylococcus aureus 7% (in French)* *Escherichia coli 6% (in %)*
Acinetobacter baumannii 20.8		
Jisha et al [39] Kerala, India (2021) *SCN 19.2%* *Escherichia coli 10.4*		
Acinetobacter baumannii 20% of the total		
Sornaranjani et al [40].	Rajiv, India (2016-2017)	*Staphylococcus aureus 16* *Staphylococcus epidermidis 16* *Escherichia coli 12*
Zelenitsky et al [45]	Winnipeg Canada (2005-2014)	*Staphylococcus epidermidis 26.5* *Streptococcus sp 12.7* *Staphylococcus aureus 9.8*
Vakilzadeh et al [46]	CHUV, Switzerland (1995-2010)	*SCN 23.9* *Streptococcus sp 14.2%* *Enterobacteria 10.6%* *Staphylococcus aureus 5.3%* *Streptococcus sp 16*
Beaudreuil et al [8]	Connecticut, United States	*SNA 34* *Staphylococcus aureus 25% of the population*
	Streptococcus sp 4.7 *SCN 29.3* RDPLF France *Staphylococcus aureus 11.6* *Streptococcus sp 8.9%*	
Ozdemir et al [47]	*SCN 30‰* *Staphylococcus aureus 17.3%* *Streptococcus sp 13.2%* Turkey (2005-2021)	

		Pseudomonas spp 6.9% *Acinetobacter* spp 6.9% *Escherichia coli* 5.5
Ajimi et al [48]	Charles-Nicolle Hospital, Tunisia (19832015)	*Staphylococcus aureus 21.7* *Staphylococcus epidermidis 12.2* *Escherichia coli 3.9*
Our study	**Sousse, Tunisia (2012-2020)**	*Staphylococcus aureus 23.4* *Escherichia coli 10.3* *Staphylococcus epidermidis 9%* *Streptococcus sp 8.3%*

V. Antibiotic resistance in the main bacteria

The problem of bacterial resistance to antibiotics particularly concerns patients with chronic renal failure (CRF). Chronic dialysis, the presence of vascular catheters, urinary catheterisation, renal transplantation, prophylactic and curative antibiotic therapy and other healthcare exposures have been identified as increased risk factors for colonisation and infection by multi-resistant bacteria. However, antibiotic resistance data concerning isolated bacteria in patients with chronic renal failure are very limited [49].

1. *Escherichia coli*

It is the most common bacterium in our study series. Belonging to group 1 of the enterobacteria, *E.coli* produces a chromosomal non-inducible cephalosporinase of the AmpC type and is sensitive to betalactam antibiotics [50]. In our study, only 17.8% of strains had a wild-type phenotype. Resistance to the combination of inhibitors and C3Gs was high compared with the studies listed in Table XXV. The rate of resistance to imipenem in our study (0.3%) is close to the result found by the Tunisian LART surveillance network (0.2%) [12]. This rate is much lower than that found by Pan et al [51], where resistance reached 35.7%.

Table XXV: Comparison of *E.coli* resistance rates to antibiotics

Authors	AMX	AMC	C3G	IMP	GN	F
Pan et al [51] (2020-2021)			57,1%	35,7		57,1%
Kissou et al [52] Burkina Faso (2019)		82,3%				44,4%
Bacha et al [53] Germany (2016)	62%	29,6%	18,3%		7%	22,4%
Hamouche et al [54] Lebanon (2009)	76,2%			30,2%	27,4%	
Michno et al [55] Poland (2015)	69,8%	50,1%	13%	0%	9%	29,6%
LART [12] Tunisia (2017)	72,4%	27,1%	19,7%	0,2%	14,2%	26,1%
ONERBA [13] France (2018)	51%	30%	11%	0%	5%	13%

Our study	82,2%	54,6%	34,3%		0,3%21,7%	47,8%

2. *Klebsiella pneumoniae*

In our study, the resistance of *K.pneumoniae* to the combination of Amoxicillin and clavulanic acid was 61%, which is higher than the data from LART in 2017 [56] and ONERBA in 2018 [13]. The highest rates of resistance were reported in the studies by Rostkowska et al. in 2018 and Cristea et al. in 2016 [57,58] (**see table XXVI**).

The rate of resistance to fluoroquinolones was 59.4%. This is relatively high compared with data from LART and ONERBA [13,56].

Table XXVI: Comparison of resistance rates of *K.pneumoniae* to antibiotics

Authors	AMC	C3G	Carbapenemes	Aminosides	FQ	SXT
Madela et al [59] South Africa (2014)	33,3%		44,4%	55,6%	77,8%	55,6%
Jukic et al [60] Bosnia-Herzegovina (2018)		46,6%	4,7%		53%	48%
Karlowsky et al. [61] United States (2002)	13,3%	11,5%		8,6%	9,7%	12,8%
Rostkowska et al. [57] Poland (2018)	73,1%	65,4%	7,7%		84,6%	61,5%
Cristea et al [58] Romania (2016)	75%	61,5%	21,4%	35,7%	35,7%	92,8%
ONERBA [13] France (2018)	32%	22%			24%	
LART [56] Tunisia 2017	41,4%	43,3%	15,9%	34,7%	35,8%	38,8%
Our study	61%	60,6%	12%	38,9%	59,4%	53,1%

3. *Staphylococcus aureus*

In the majority of the studies mentioned in Table XXVII, the resistance of *S.aureus* resistance to penicillin G exceeds 80%. A study carried out in France reported a decrease in penicillin G resistance over the last decade [62]. Recent observations in North America [63] and Sweden [64] report sensitivity rates of *Staphylococcus aureus* to penicillin G of up to 30%.

Table XXVII: Comparison of resistance rates of *Staphylococcus aureus* to antibiotics

Authors	PG	OXA	Aminosides	ERY	FQ	Glycopep
Khurana et al [65] India (2012-2016)	99,1%	55,8%	37,8	62,7%	66,6%	1,2%

33

Gitau et al [66] Kenya (2014-2016)	92%	29%	13%		26%	22%	3%
Kitara et al [67] Oughanda (2011)	81,5%	2,6%	0%		10,5%	2,6%	
Fortuin-de Smidt [68] South Africa (2012-2013)		36%	45,8%	41,7		37,9%	2%
Hassan et al [69] Egypt (2018-2020)	94,2%		42,3%		17,3%	69,2%	3,8%
Wang et al [42] Taiwan (2007-2016)	73,3%	13,3%				7,1%	
ONERBA [13] France 2018		13,4%	2,3%		28,9%	14%	
LART [70] Tunisia 2017	94,7%	19,3%	29,6%		17,4%	10,2%	0%
Our study (2012-2020)	89,7%	13%	13,3%		11,8%	53,1%	1,2%

4. *Pseudomonas aeruginosa*

In our series, the rate of resistance of our *P.aeruginosa* isolates to the piperacillin-tazobactam combination (15.3%) is close to the rate reported by the Tunisian surveillance network (18.5%) [71]. Resistance to ceftazidime (14.4%) is slightly lower than that reported by the LART and ONERBA surveillance networks [13,71]. The highest rates of resistance were reported by Gill et al. in Pakistan in 2010 and Krir et al. in Ben Arous in 2018 [72]. The rate of resistance to ciprofloxacin (27.7%) is slightly higher than the data from LART and ONERBA but remains lower than the rate of resistance reported by a study carried out in Egypt in 2007, which was 29% [73]. The rate of resistance to amikacin (7.9%) found in our study was lower than that reported by LART and ONERBA. All our strains were sensitive to colistin (**see Table XXVIII**).

Table XXVIII: Comparison of resistance rates of *Pseudomonas aeruginosa* to antibiotics

Authors	PIPTZ	CAZ	IMP	CIP	AMK	CT
Gill et al [74] Pakistan (2010)	36,6%	70,7%	90,2%	87,8%	83%	31,7%
Ndip et al [75] Cameroon (2014)		13,7%		0%	5,9%	
Gad et al [73] Egypt (2007)				29%	8%	
Krir et al [72] Ben Arous (2012 2018)	57,8%	35,7%	63,2%	42,9%	68,9%	0,8%
Kpoda et al [76] Burkina Faso (2017-2018)		23,5%				0%0%

ONERBA [13] France 2018		18%	18%	17%	12%	
LART [71] Tunisia 2017	18,5%	16,1%	19,9%	19,6%	15,6%	
Our study	15,3%	14,4%	8%	27,7%	7,9%	0%

5. *Acinetobacter baumannii*

It is the second non-fermenting Gram-negative bacillus found (n= 38, 1.3%) in our study. The resistance rates of our strains to the various antibiotics tested exceeded 50%. Our results are in line with those found by Lob et al [77] in the different continents of the world, with the exception of North America where resistance rates are much lower, mainly to ceftazidime, imipenem, the piperacillin-tazobactam combination and amikacin. All our strains were sensitive to colistin. Mellouli et al [78] found a colistin resistance rate of 1.8%.

Table XXIX: Comparison of resistance rates of *Acinetobacter baumannii* to antibiotics

		PIP	TZ	CAZ	IMP	CIP	AMK
Lob et al. [77] (2013-2014)	**Africa**	83,1%	72,3%	81,5%	72,3%	53,3%	
	Europe	88,6%	92,5%	88,6%	95,5%	80,1%	
	Middle East	96,7%	94,1%	91,4%	96,1%	67,8%	
	Asia	78%	78,3%	73,1%	78,8%	68,1	
	South America	42,5%	53,2%	36,2%	68,1%	38,3%	
	North America						
	Latin	87,2%	85,1%	83,7%	90,1%	80,8%	
Our study		72,4%	79,4%	62,5%	82,5%	52,5%	

VI. Multi-resistant bacteria

Bacterial multiresistance is a major public health problem. In fragile patients such as those in nephrology, the acquisition of a BMR increases the rate of morbidity and mortality and risks leading to a therapeutic impasse.

One study showed that stages G2 to G5 of CKD are an important risk factor for overall antimicrobial resistance, and more importantly, for multiple drug resistance, with CKD increasing the risk of BMR infections by approximately twofold. Patients on chronic hemodialysis had a four-fold increased risk of acquiring BMR infections [79].

In our study, 688 of the isolates were BMR, i.e. 24.1%. Majeed et al. reported an incidence of 42.9% of BMR in patients with renal disease [80].

The rate of multi-resistant BGN in our study was 21.5% (n=614). A study carried out in an outpatient hemodialysis facility in the United States revealed that 28% of patients were colonised by a BGN resistant to at least three of the six antibiotics tested. Furthermore, 20% of patients were colonised by one of these multi-resistant BGNs during a 6-month follow-up period [81].

Despite the absence of large-scale epidemiological studies on multi-resistant BGN infection in nephrology patients, the frequent exposure of these patients to antibiotics and regular hospital stays increase their risk of colonisation by these multi-resistant bacteria. Available data suggests that these multi-resistant bacteria are relatively common in dialysis patients.

In our study, 457 strains of enterobacteria (31.2% of all enterobacteria) were resistant to C3G. The number of C3G-resistant *K.pneumoniae* strains was 190 (40.7%). The number of C3G-resistant *Escherichia coli* was 235 (32.1%). These figures are high compared with data from North American and European studies, which report a C3G resistance rate of 27.2% for *Klebsiella pneumoniae* and 22.1% for *Escherichia coli* [82].

Statistics from BMR surveillance networks indicate that *Klebsiella pneumoniae* and *Escherichia coli* are the two enterobacteria whose frequency has been rising steadily over the last 10 years [83]. A number of studies have focused on the role of antibiotics in the emergence of C3G-resistant Enterobacteriaceae, in addition to the acquisition of new virulence factors by these germs.

In our study, 126 isolated strains were carbapenem-resistant enterobacteria (8.7%). This may be due to overuse of this class of antibiotics in the department, in addition to predisposing factors, mainly catheterisation. In 2019, the CDC *(Centers for Disease Control and Prevention)* estimated that carbapenem-resistant enterobacteria are responsible for 13,100 infections and 1,100 deaths in the United States each year [84].**See table XXX**.

In 2014, 10.9% of catheter-related infections were due to carbapenem-resistant *Klebsiella* spp and 1.9% were due to carbapenem-resistant *E.coli* [85].

The Reseau d'Alerte d'Investigation et de Surveillance des Infections Nosocomiales (RAISIN) has reported a steady decline in MRSA in favour of C3G-resistant enterobacteria since 2006 [83]. In our study, 49 isolated strains were MRSA, i.e. 12.9% of all isolated *S.aureus* strains. Compared with the general population, patients with kidney disease are more likely to be affected by both meticillin-susceptible and meticillin-resistant *S. aureus* infection. This high susceptibility is associated with greater morbidity and mortality in nephrology patients. Reported rates of MRSA colonisation in hemodialysis patients range from 2.3% to 27.3% [86]. It should be noted that up to 35% of colonised patients develop MRSA infections within a year [87]. The US Centers for Disease Control and Prevention (CDC) have reported that invasive MRSA infections affect more than 4 in 100 dialysis patients, an incidence more than 100 times higher than that seen in the general population [88]. In renal transplant patients, MRSA colonisation varies from 1.2% to 12.5% [89]. In 2017, the report of the Algerian Network on Antimicrobial Resistance indicated that approximately 23% of *S.aureus* were resistant to meticillin [90].

In our study, 17 strains of *Pseudomonas* sp (15.2%) were resistant to ceftazidime. For *P.aeruginosa,* 14 strains (14.4%) were resistant to ceftazidime. A multicentre study found a prevalence of 12.4% of multi-resistant *P.aeruginosa* in Iraq, 54% in Syria, 64.5% in Lebanon, 47.6% in Palestine, 52.5% in Jordan, 7.3% in Saudi Arabia, 75.6% in Egypt and 54% in Tunisia [91].

In the case of *A.baumannii,* 25 strains resistant to imipenem were isolated (62.6%). Resistance to imipenem alone classifies this bacterium as a BMR. The prevalence of ABRI is constantly increasing worldwide [92]. The spread of ABRI could be via contaminated equipment or the hands of healthcare staff [93]. Higgins et al. reported that 95.5% of *Acinetobacter baumannii* strains were ABRI [94].

In the case of enterococci, 25 glycopeptide-resistant strains (8.2%) were identified. *Enterococcus faecalis* and *Enterococcus faecium* were the two most isolated species in our study. In fact, these are the two species most implicated in enterococcal infections in humans[95]. In 1980, the first acquired resistance to vancomycin in enterococci was described, particularly in *E. faecium* [95]. In patients with chronic renal failure, colonisation and infection with glycopeptide-resistant enterococci is frequent. Several studies around the world have reported an ERG colonisation rate in hemodialysis patients ranging from 2.8% to 10.8% [96]. Recent hospitalisation and previous antibiotic use are risk factors for colonisation [97]. In a 2014 US study of hemodialysis patients, 11.4% of enterococcal isolates were resistant to glycopeptides [98].

Table XXX: **Comparison of BMRs in different studies**

	CDC (2019) [84]	Majeed et al [80]	Weiner and al.[82]	Our study
EBRC3G	N=97400 9100 deaths			**32,2%**
E.coli-R-C3G		95%	22,2%	**32,1%**
K.p-R- **C3G**		82,3%	24,1%	**40,7%**
ERC	N=13100 1100 ddces			**8,7%**
E.coli- **R -C**			14,1%	
K.p- **R-C**			10,9%	
Pseudo-Caz-R	N=32600 2700 deaths	80%	17,9%	**15,2%**
SHELTER	N= 8500		43,7%	**62,6%**

	700 deaths	
BGN BMR TOTAL	42,9%	21,5%
MRSA	50,7%	
ERG	46%	

EBRC3G: Enterobacteria resistant to C3G; E.coli-R-C3G: E.coli resistant to C3G; K.p-R-C3G: K.pneumoniae resistant to C3G; E.coli-R-C: E.coli resistant to carbapenemes; K.p-R-C: Carbapenem-resistant K.pneumoniae; Pseudo-CAZ-R: Ceftazidime-resistant Pseudomonas spp; BGN BMR: Multi-resistant Gram-negative bacilli

In our study, we compared the difference in BMR prevalence between the "before 2016" period and the "after 2016" period. It was statistically significant for the following BMRs: C3G-resistant enterobacteria, glycopeptide-resistant enterococci, MRSA and ceftazidime-resistant *P.aeruginosa*. This could be explained by the increased shortage of staff during the "post-2016" period. Since 2015, a number of paramedical staff have left without being replaced. The workload has therefore increased, which has probably revealed some shortcomings in hygiene measures. In addition, the nephrology department was restructured during this period. In the past, there were three wards on the women's side, each with 4 beds. Then, one of the wards was converted into a transplant unit. As a result, the women's ward has become very cramped. It comprises just two wards with 6 beds in each room. Isolating patients with BMR infections is therefore impossible from a practical point of view. This proximity to patients, combined with a lack of staff, could therefore explain the increase in the prevalence of BMR during the "post-2016" period.

VII. Self-criticism of the study

Our study has a number of limitations that deserve to be emphasised. Firstly, this was a retrospective study, and any missing data could not be recovered, nor could any additional examinations be carried out. The lack of clinical information for certain samples meant that we were unable to distinguish between infection, colonisation, carriage or contamination. Due to an archiving problem, we were unable to complete the figures for the annual incidence of infections in the nephrology department for the period from 2012 to 2016.

It should also be noted that there are few articles on bacterial infections in the nephrology department. This problem forced us to compare our results with those of patients hospitalised in the various clinical departments, including those in intensive care.

5 CONCLUSION

Nephrology patients with kidney disease (chronic renal failure, dialysis patients, transplant patients) have a high risk of developing a bacterial infection compared with the general population. Such infections can rapidly lead to serious septic states that can be life-threatening (peritonitis, bacteremia with risk of endocarditis). Nowadays, bacterial resistance is a major public health problem, especially as these patients are immunocompromised and wear devices such as catheters. This resistance is constantly increasing and is responsible for considerable morbidity and mortality.

The aims of this study are to identify the germs responsible for bacterial infections in the nephrology department, to study the resistance profiles of these germs to the various antibiotics and to determine the frequency of bacterial infections in dialysis patients (hemodialysis, peritoneal dialysis) in order to improve the management of nephrology patients by adapting probabilistic antibiotic therapy according to the bacterial ecology of the department.

We carried out a retrospective descriptive and analytical study at the Sahloul University Hospital Centre on all non-redundant bacterial strains isolated from samples sent to the microbiology laboratory by the nephrology hospitalization unit, the hemodialysis unit and the continuous peritoneal dialysis (CPD) unit, over a period of 9 years, from 1 January 2012 to 31 December 2020. Bacterial identification was carried out using conventional and automated bacteriological methods. Bacterial identification was performed using conventional and automated bacteriological methods, and antibiotic susceptibility testing of isolates was carried out in accordance with CA-SFM/EUCAST recommendations.

The Chi-square test was used to compare trends in the prevalence of BMRs in the nephrology department between the two periods: "before 2016" and "after 2016".

During the study period, 2851 isolates were identified from the nephrology department, the majority of which belonged to the nephrology inpatient unit (n=2363, 83%), followed by the DPC unit (n=264, 9.2%) and the HD unit (n=224, 7.8%).

At the nephrology inpatient unit, urinary tract infections were in the majority (n=1762, 74.6%), followed by bacteremia (N=308, 13%). *Escherichia coli* (N=1762, 74.6%) and *Klebsiella pneumoniae* (n=501, 21.2%) were the most frequently found uropathogenic bacteria. *Staphylococcus aureus* was the most isolated bacterium in bacteremia (n=140, 45.4%).

At the hemodialysis unit, urinary tract infections (n=84, 37.5%) and bacteremia (n=83, 37%) were in the majority. *Staphylococcus aureus* was the main germ found (n=87, 38.9%).

In the CPB unit, peritonitis was the most frequent infection (n= 145, 54.9%) and *Staphylococcus aureus* was the main germ found (n=34, 23.4%). Although peritoneal dialysis is the least invasive method, the complications caused by this method of purification are not negligible and infectious peritonitis is the main complication.

Escherichia coli was the most common bacterium in our study. More than half the strains were resistant to amoxicillin (82.2%) and the combination of amoxicillin and clavulanic acid (54.6%). Resistance to C3G (34.3%) and fluoroquinolones (47.8%) was also high. Only 0.3% of strains were resistant to imipenem and all strains were sensitive to colistin.

The resistance rates of *Klebsiella pneumoniae* strains to C3Gs and carbapenems were 60.6% and 12% respectively. Resistance to fluoroquinolones in these isolates was high. It reached 59.4%.

Around 90% of *Staphylococcus aureus* strains produced penicillinase, 13% were resistant to meticillin and 1.2% were resistant to glycopeptides.

BMR are a serious problem in our department, in the hospital and even worldwide. Among the 2851 strains isolated in nephrology, 688 multi-resistant bacteria were identified, representing 24.1% of isolates. C3G-resistant enterobacteria were the most common type of MRB (n=457, 16%), followed by carbapenem-resistant enterobacteria (n=126, 4.4%). In third place was MRSA (n=49, 1.73%), followed by glycopeptide-resistant enterococci (n=25, 0.9%), imipenem-resistant *A.baumannii* (n=16, 0.6%) and ceftazidime-resistant *Pseudomonas* spp (n=15, 0.5%).

The emergence of resistant bacteria should be controlled in order to slow their growth and prevent their spread to the community environment, as well as the transfer of resistance genes to other bacteria. Controlling the use of antibiotics by clinicians and epidemiological monitoring of resistance by the microbiology laboratory will help to bring bacterial resistance under control. A collective effort via local surveillance of antibiotic resistance, supervised by a national network such as LART, is desirable in order to provide more accurate data on the evolution of bacterial resistance in this category of patient.

The prevention of catheter-related infections has recently been updated: the use of sterile gloves, masks and sterile gowns when inserting the catheter, the use of skin antiseptics, and compliance with protocols for disinfecting dialysis stations, patients and staff. All these measures significantly reduce the risk of developing a catheter-related infection.

In order to prevent bacterial infections in patients admitted to the nephrology department, it is recommended that hygiene and asepsis measures be strengthened (individual and collective cleanliness, maintenance of the establishment) and that antibiotics be prescribed rationally and appropriately, remaining the best means of managing this problem. Educating patients by explaining their illness and the importance of adhering to treatment will partly reduce the risk of bacterial infection.

Recommendations & Outlook

In our study, 688 multi-resistant bacteria (24.1%) were isolated, of which 614 were BGN. In order to prevent BMR infections in nephrology patients, hygiene measures must be strengthened by continuously stocking up on soap, hydroalcoholic gel, antiseptic solutions and protective equipment (sterile gloves, masks, gowns, etc.). It is also necessary to disseminate national and international recommendations and raise awareness among department staff of the impact of compliance with asepsis measures (washing hands before and after any care procedure, disinfection with a product approved for hospital use, recognised as effective and homologated) on reducing the incidence of nosocomial infections. Patients also need to be educated about their illness and the hygiene measures to be taken to reduce the risk of infection. Access to hemodialysis rooms should be restricted to all foreigners, including those accompanying patients. Systematic screening for the carriage of BMRs in these patients on admission and on a weekly basis is recommended. In the HD and DPC units, where the most frequent BMR are gram-positive cocci, screening for MRSA using nasal, axillary and buccal swabs is recommended. In the nephrology unit, where multi-resistant BGNs were the most common, rectal swabs were recommended to detect them. Geographical and technical isolation of patients with a BMR infection

is required since the genetic basis of multidrug resistance is a mobile gene (plasmids, transposants) [99, 100].

In the future, we would like to repeat a comparative study to assess the effectiveness of BMR screening in reducing BMR infections in the nephrology department.

Inserting this catheter is an invasive procedure. This procedure requires compliance with asepsis and hygiene measures (hand washing, sterile gloves, use of antiseptic solutions) and antibiotic prophylaxis during catheter insertion could be proposed[100]. Consideration should also be given to setting up good practice guidelines for catheter insertion in the department, accessible to staff, and maintaining close collaboration between doctors and nurses.

Probabilistic antibiotic therapy should cover the most frequently isolated germs in each infectious site, taking into account the resistance rates found in our work. For urinary tract infections, which are in first place in our study, *E.coli* and *K. pneumoniae* are the most isolated bacteria. It is recommended to use an antibiotic with a resistance rate of <10%[102]. For lower urinary tract infections, fosfomycin may be recommended (resistance rate 1.9% for *E.coli* and 5.8% for *K.pneumoniae)*. This antibiotic is known for its predominant urinary elimination in active form, its low cost and its low potential for selecting resistant bacteria[102]. For acute pyelonephritis, carbapenems should be preferred to C3Gs, since in our study the rate of resistance to C3Gs was 34.3% for *E. coli* and 60.6% for *K. pneumoniae*. Given the risk of selection of resistant mutants when using this class of antibiotic, a re-evaluation of the microbiological results is necessary within the first 48 hours of its introduction as a probabilistic treatment[103]. At the nephrology inpatient unit, bacteremia is due to *S.aureus* followed by *E.coli*. It is therefore necessary to cover both Gram-positive and Gram-negative bacteria. A combination of imipenem + vancomycin is therefore recommended. In peritoneal dialysis patients, bacterial peritonitis was the first complication found. *S.aureus* was the most isolated bacterium, followed by *E.coli*. In 2016, the ISPD (*International Society for Peritoneal Dialysis*) recommended covering both Gram-positive bacteria (with vancomycin or C1G) and BGN (with C3G or an aminoglycoside)[104]. Sensitivity statistics show excellent antibiotic coverage of Gram-positive bacteria with a glycopeptide and Gram-negative bacteria with a combination of C3G and an aminoglycoside, which was also recommended by Beaudreuil et al[105].

In order to reduce the rate of antibiotic resistance in our department and throughout the hospital, we urgently need to set up a team of antibiotic therapy referents to provide advice and raise awareness in all departments about appropriate and rational antibiotic therapy. This team will closely monitor antibiotic prescriptions, with the aim of reducing resistance and the cost of care.

6 REFERENCES

1. **Stel VS, van de Luijtgaarden MW, Wanner C, Jager KJ; on behalf of the European Renal Registry Investigators.** The 2008 ERA-EDTA Registry Annual Report-a precis. *NDTPlus 2011;4:1-13.*

2. **GPR website.** Infectiology and renal failure. *[On line]. 2020 [Accessed 21/08/2022], available at URL: http://sitegpr.com/fr/rein/en-savoir-plus/infectiologie-et-insuffisance-renale/*

3. **Johnson DW, Fleming SJ.** The use of vaccines in renal failure. *Clin Pharmacokinet 1992;22:434-46.*

4. **Monnet DL.** Antibiotic consumption and bacterial resistance. *Ann Fr Anesth Reanim 2000;19:409-17.*

5. **Magiorakos AP, Srinivasan A, Carey RB, Carmeli Y, Falagas ME, Giske CG, et al.** Multidrug-resistant, extensively drug-resistant and pandrugresistant bacteria: an international expert proposal for interim standard definitions for acquired resistance. *Clin Microbiol Infect 2012;18:268-81.*

6. **Parker CM, Kutsogiannis J, Muscedere J, Cook D, Dodek P, Day AG, et al.** Ventilator-associated pneumonia caused by multidrug-resistant organisms or *Pseudomonas aeruginosa*: Prevalence, incidence, risk factors, and outcomes. *JCrit Care 2008;23:18-26.*

7. **Societe Frangaise de Microbiologie.** CA-SFM/EUCAST April 2021 V1.0. *Paris: SFM; 2021.*

8. **Beaudreuil S, Hebibi H, Charpentier B, Durrbachr A.** Serious infections in peritoneal dialysis and chronic conventional hemodialysis patients: peritonitis and vascular access infections. *Reanimation 2008;17:233-41.*

9. **Meurice L, Vilain P, Maillard L, Revel P, Caserio-Schonemann C, Filleul L.** Impact des deux confinements sur le recours aux soins d'urgence lors de l'epidemie de COVID-19 en Nouvelle-Aquitaine. *Sante Publique 2021;33:393-7.*

10. **Heist T, Schwartz K, Butler S.** Trends in overall and non-COVID-19 hospital admissions. *[On-line]. 2021 [Accessed 21/08/2022], available at URL: https://www.kff.org/health-costs/issue-brief/trends-in-overall-and- non-covid-19-hospital-admissions/*

11. **Samanipour A, Dashti-Khavidaki S, Abbasi MR, Abdollahi A.** Antibiotic resistance patterns of microorganisms isolated from nephrology and kidney transplant wards of a referral academic hospital. *J Res Pharm Pract 2016;5:43.*

12. **LART.** Data from 2017: *E. coli* (N = 8266). *[Online]. 2018 jConsuhe on 21/08/2022], availablea the URL: https://www.infectiologie.org.tn/pdf_ppt_docs/resistance/1544218181.pdf*

13. **ONERBA.** Activity report. May 2020 edition. *[Online]. 2018 [Accessed*

21/08/2022], available at the URL   :http://onerba-doc.onerba.org/Reports/Rapport-ONERBA- 2018/Rap18_onerba_synthese.pdf

14. **Bidet P, Bonarcorsi S, Bingen E.** Pathogenic factors and pathophysiology of extraintestinal *Escherichia coli. Arch Pediatr 2012;19:S80-92.*

15. **Wang Y, Li H, Chen B.** Pathogen distribution and drug resistance of nephrology patients with urinary tract infections. *Saudi Pharm J 2016;24:337-40.*

16. **Shrimali G, Patel K.** Bacteriology and antibiogram of urinary tract infection of chronic renal failure patients taking hemodialysis at tertiary care centre. *Saudi J Pathol Microbiol 2019;4:175-8.*

17. **Adhikari L.** Incidence of bacterial infections in chronic kidney disease patients admitted in nephrology unit of Kathmandu Medical College Teaching Hospital. *J Kathmandu Med Coll 2018;7:26-9.*

18. **Tayh G, Al Laham N, Ben Yahia H, Ben Sallem R, Elottol AE, Ben Slama K.** Extended-spectrum в -lactamases among Enterobacteriaceae isolated from urinary tract infections in Gaza Strip, Palestine. *BioMed Res Int 2019;2019: 4041801.*

19. **Mawufemo TY, Dzidzonu NK, Deassoua BL, Georges TK, Badomta D, Majeste WI.** Bacteriological profile of urinary tract infections in patients with chronic renal failure hospitalised at the nephrology department. *J Rech Sci Univ Lome 2020;22:237-41.*

20. **Chemlal A, Karimi I, Benabdellah N, Alaoui F, Alaoui S, Haddiya I, et al.** Urinary tract infections in patients with chronic renal failure in nephrology: bacteriological profile and prognosis. *Nephrol Ther 2015;11:399.*

21. **Guermazi-Toumi S, Boujlel S, Assoudi M, Issaoui R, Tlili S, Hlaiem ME.** Susceptibility profiles of bacteria causing urinary tract infections in Southern Tunisia. *J Glob Antimicrob Resist 2018;12:48-52.*

22. **Lagier JC, Letranchant L, Selton-Suty C, Nloga J, Aissa N, Alauzet C, et al.** *Staphylococcus aureus* bacteremia and endocarditis. *Ann Cardiol Angeiol 2008;57:71-7.*

23. **Vandenbos F, Ozanne A, Burel-Vandenbos F, Tempesta S, Fosse T, Dellamonica P.** *Escherichia coli* bacteria of urinary origin. *Presse Med 2004;33:847-51.*

24. **Rosenthal VD, Belkebir S, Zand F, Afeef M, Tanzi VL, Al-Abdely HM, et al.** Six-year multicenter study on short-term peripheral venous catheters-related bloodstream infection rates in 246 intensive units of 83 hospitals in 52 cities of 14 countries of Middle East: Bahrain, Egypt, Iran, Jordan, Kingdom of Saudi Arabia, Kuwait, Lebanon, Morocco, Pakistan, Palestine, Sudan, Tunisia, Turkey, and United Arab Emirates - International Nosocomial Infection Control

Consortium (INICC) findings. *J Infect Public Health 2020;13:1134-41.*

25. **Mohamed H, Ali A, Browne LD, O'Connell NH, Casserly L, Stack AG, et al.** Determinants and outcomes of access-related blood-stream infections among Irish haemodialysis patients; a cohort study. *BMC Nephrol 2019;20:68.*

26. **Agrawal V, Valson AT, Mohapatra A, David VG, Alexander S, Jacob S, et al.** Fast and furious: a retrospective study of catheter-associated bloodstream infections with internal jugular nontunneled hemodialysis catheters at a tropical center. *Clin Kidney J2019;12:737-44.*

27. **Zhang HH, Cortes-Penfield NW, Mandayam S, Niu J, Atmar RL, Wu E, et al.** Dialysis catheter-related bloodstream infections in patients receiving hemodialysis on an emergency-only basis: a retrospective cohort analysis. *Clin Infect Dis 2019;68:1011 -6.*

28. **Fram D, Okuno MFP, Taminato M, Ponzio V, Manfredi SR, Grothe C, et al.** Risk factors for bloodstream infection in patients at a Brazilian hemodialysis center: a case-control study. *BMC Infect Dis 2015;15:158.*

29. **Alhazmi SM, Noor SO, Alshamrani MM, Farahat FM.** Bloodstream infection at hemodialysis facilities in Jeddah: a medical record review. *Ann Saudi Med 2019;39:258-64.*

30. **Yamashita K, Ishiyama Y, Yoshino M, Tachibana H, Toki D, Konda R, et al.** Urinary tract infection in hemodialysis-dependent end-stage renal disease patients. *Res Rep Urol 2022;14:7-15.*

31. **Manhal FS, Mohammed AA, Ali KH.** Urinary tract infection in Hemodialysis patients with renal failure. *J Fac Med Baghdad 2012;54:38-41.*

32. **Rahman MS, Ahmed MU, Mahtab Uddin BM, Chowdhury RK, Ur Rasid H, Begum A.** Bacteriological profile and their antimicrobial susceptibility pattern of urinary tract infection patients attending at the nephrology department of Enam Medical College and Hospital, Savar, Dhaka. *Med Today 2022;34:51-6.*

33. **Fisher M, Golestaneh L, Allon M, Abreo K, Mokrzycki MH.** Prevention of Bloodstream Infections in Patients Undergoing Hemodialysis. *Clin J Am Soc Nephrol 2020;15:132-51.*

34. **Bhojaraja MV, Prabhu RA, Nagaraju SP, Rao IR, Shenoy SV, Rangaswamy D, et al.** Hemodialysis catheter-related bloodstream infections: a single-center experience. *J Nephropharmacology 2022;11:e10475.*

35. **Abd El-Hamid El-Kady R, Waggas D, AkL A.** Microbial repercussion on hemodialysis catheter-related bloodstream infection outcome: a 2-year retrospective study. *Infect Drug Resist 2021;14:4067-75.*

36. **Krishnan A, Irani K, Swaminathan R, Boan P.** A retrospective study of tunnelled haemodialysis central line-associated bloodstream infections. *J*

Chemother 2019;31:132-6.

37. **Lungren MP, Christensen D, Kankotia R, Falk I, Paxton BE, Kim CY.** Bacteriophage K for reduction of *Staphylococcus aureus* biofilm on central venous catheter material. *Bacteriophage 2013;3:e26825.*

38. **Shanks RM, Sargent JL, Martinez RM, Graber ML, O'Toole GA.** Catheter lock solutions influence staphylococcal biofilm formation on abiotic surfaces. *Nephrol Dial Transplant 2006;21:2247-55.*

39. **Jisha TU, Sarada Devi KL.** A clinico-microbiological profile of peritonitis in continuous ambulatory peritoneal dialysis patients. *Int J Curr Microbiol Appl Sci 2021;10:914-21.*

40. **Sornaranjani M, Ramani C, Karuppasamy K.** Assessment of microbiological profile in peritoneal dialysis in patients with chronic kidney disease. *Panacea J Med Sci 2022;12:305-10.*

41. **Ghali JR, Bannister KM, Brown FG, Rosman JB, Wiggins KJ, Johnson DW, et al.** Microbiology and outcomes of peritonitis in Australian peritoneal dialysis patients. *Perit Dial Int JInt Soc Perit Dial 2011;31:651 -62.*

42. **Wang HH, Huang CH, Kuo MC, Lin SY, Hsu CH, Lee CY, et al.** Microbiology of peritoneal dialysis-related infection and factors of refractory peritoneal dialysis related peritonitis: A ten-year single-center study in Taiwan. *J Microbiol Immunol Infect 2019;52:752-9.*

43. **Song P, Yang D, Li J, Zhuo N, Fu X, Zhang L, et al.** Microbiology and Outcome of Peritoneal Dialysis-Related Peritonitis in Elderly Patients: A Retrospective Study in China. *Front Med (Lausanne) 2022;9:799110.*

44. **Kanjanabuch T, Chatsuwan T, Udomsantisuk N, Nopsopon T, Puapatanakul P, Halue G, et al.** Association of local unit sampling and microbiology laboratory culture practices with the ability to identify causative pathogens in peritoneal dialysis-associated peritonitis in Thailand. *Kidney Int Rep 2021;6:1118-29.*

45. **Zelenitsky SA, Howarth J, Lagace-Wiens P, Sathianathan C, Ariano R, Davis C, et al.** Microbiological trends and antimicrobial resistance in peritoneal dialysis-related peritonitis, 2005 to 2014. *Perit Dial Int J Int Soc Perit Dial 2017;37:170-6.*

46. **Vakilzadeh N, Burnier M, Halabi G.** Infectious peritonitis in peritoneal dialysis: an overly dreaded complication *Rev Med Suisse 2013;9:446- 50.*

47. **Ozdemir A, Yucel Kogak S.** Peritoneal dialysis-related peritonitis: microbiological profile and outcome. *Bakirkoy Tip Derg Med J Bakirkoy 2022;18:25-30.*

48. **Ajimi K, Barbouch S, Najjar M, Ounissi M, Ben Hmida F, Harzallah A, et al.** Peritonitis and peritoneal dialysis. *Nephrol Ther 2021;17:371.*

49. **Wang TZ, Kodiyanplakkal RP, Calfee DP.** Antimicrobial resistance in nephrology. *Nat Rev Nephrol 2019;15:463 -81.*

50. **Courvalin P, Leclercq R, Bingen E.** Antibiogramme. 3^{cm} th edition. *Paris: ESKA; 2012.*

51. **Pan D, Peng P, Fang Y, Lu J, Fang M.** Distribution and drug resistance of pathogenic bacteria and prognosis in patients with septicemia bloodstream infection with renal insufficiency. *Infect Drug Resist 2022;15:4109-16.*

52. **Kissou PF, Semde A, Sawadogo A, Tao M, Sanou G, Coulibaly G.** Antibiotic resistance of urinary tract infections in the nephrology-dialysis department of the Souro Sanou University Hospital in Bobo-Dioulasso (Burkina Faso). *Kidney Int Rep 2022;7:S69.*

53. **Abo Basha J, Kiel M, Gorlich D, Schutte-Nutgen K, Witten A, Pavenstadt H, et al.** Phenotypic and genotypic characterization of escherichia coli causing urinary tract infections in kidney-transplanted patients. *J Clin Med 2019;8:988.*

54. **Hamouche E, Sarkis DK.** Evolution of antibiotic susceptibility of *Escherichia coli, Klebsiella pneumoniae, Pseudomonas aeruginosa* and *Acinetobacter baumanii* in a Beirut University Hospital between 2005 and 2009. *Pathol Biol 2012;60:e15-20.*

55. **Michno M, Sydor A, Walaszek M, Sulowicz W.** Microbiology and drug resistance of pathogens in patients hospitalized at the nephrology department in the South of Poland. *Pol J Microbiol 2018;67:517-24.*

56. **LART.** 2017 Data: *K. pneumoniae* (N=3178). *[Online]. 2018 [Accessed 21/08/2022], Available at lURL:*
https://www.infectiologie. org.tn/pdf_ppt_docs/resistance/1544218300.pdf

57. **Rostkowska OM, Kuthan R, Burban A, Salinska J, Ciebiera M, MIynarczyk G, et al.** Analysis of susceptibility to selected antibiotics in klebsiella pneumoniae, escherichia coli, enterococcus faecalis and enterococcus faecium causing urinary tract infections in kidney transplant recipients over 8 years: single-center study. *Antibiotics 2020;9:284.*

58. **Cristea OM, Avramescu CS.** Urinary tract infection with *Klebsiella pneumoniae* in patients with chronic kidney disease. *Curr Health Sci J 2017;2:137-48.*

59. **Madela Y.** The incidence and antimicrobial sensitivity of urinary tract infections in HIV positive and negative nephrology patients at Inkosi Albert Luthuli Central Hospital in Kwazulu Natal Province, South Africa [Thesis] *Durban: School of Nelson Rholihlahla Mandela School of Medicine, University of KwaZulu Natal; 2018.*

60. **Jukic I, Topic D, Aculic EJ, Dedeic-Ljubovic A.** Frequency and

antimicrobial susceptibility pattern of hospital isolates of *Escherichia coli* and *Klebsiella pneumoniae* in urine samples. *Acta Medica Salin 2019;49:195- 200.*

61. Karlowsky JA, Jones ME, Draghi DC, Thornsberry C, Sahm DF, Volturo GA. Prevalence and antimicrobial susceptibilities of bacteria isolated from blood cultures of hospitalized patients in the United States in 2002. *Ann Clin Microbiol Antimicrob 2004;3:7.*

62. Moraly J, Dahoumane R, Dubee V, Preda G, Baudel JL, Joffre J, et al. Penicillin G susceptibility in *Staphylococcus aureus* is not so infrequent. *Minerva Anestesiol 2018;84:123-24.*

63. Chabot MR, Stefan MS, Friderici J, Schimmel J, Larioza J. Reappearance and treatment of penicillin-susceptible *Staphylococcus aureus* in a tertiary medical centre. *JAntimicrob Chemother 2015;70:3353-6.*

64. Resman F, Thegerstrom J, Mansson F, Ahl J, Tham J, Riesbeck K. The prevalence, population structure and screening test specificity of penicillin-susceptible *Staphylococcus aureus* bacteremia isolates in Malmo, Sweden. *J Infect 2016;73:129-35.*

65. Khurana S, Mathur P, Malhotra R. *Staphylococcus aureus* at an Indian tertiary hospital: Antimicrobial susceptibility and minimum inhibitory concentration (MIC) creep of antimicrobial agents. *J Glob Antimicrob Resist 2019;17:98-102.*

66. Gitau W, Masika M, Musyoki M, Museve B, Mutwiri T. Antimicrobial susceptibility pattern of Staphylococcus aureus isolates from clinical specimens at Kenyatta National Hospital. *BMC Res Notes 2018;11:226.*

67. Kitara L, Anywar A, Acullu D, Odongo-Aginya E, Aloyo J, Fendu M. Antibiotic susceptibility of *Staphylococcus aureus* in suppurative lesions in Lacor Hospital, Uganda. *Afr Health Sci 2011;11:34-9.*

68. Fortuin-de Smidt MC, Singh-Moodley A, Badat R, Quan V, Kularatne R, Nana T, et al. *Staphylococcus aureus* bacteraemia in Gauteng academic hospitals, South Africa. *Int J Infect Dis 2015;30:41 -8.*

69. Hassan EA, Abdel-Rahim MH, Mohamed TH, Azoz NMA, Mohamed ME. Some virulence genes of *Staphylococcus aureus* isolated from infected vascular accesses in hemodialysis patients at Assiut University Hospitals. *Bull Pharm Sci Assiut 2022;45:371-87.*

70. LART. Data from 2017: *S'. aureus* (N=1790). *[Online]. 2018 [Consulate on 21/08/2022], available at URL:*

https://www.infectiologie.org.tn/pdf_ppt_docs/resistance/1544218573. pdf

71. LART. 2017 Data: *P. aeruginosa* (N=1968). *[Online]. 2018 [Accessed 21/08/2022], Available from URL:*

https://www.infectiologie.org.tn/pdf_ppt_docs/resistance/1544218416. pdf

72. **Krir A, Dhraief S, Messadi AA, Thabet L.** Bacteriological profile and antibiotic resistance of bacteria isolated in a burns resuscitation unit over seven years. *Ann Burns Fire Disasters 2019;32:197-202.*

73. **Gad GF, El-Domany RA, Zaki S, Ashour HM.** Characterization of *Pseudomonas aeruginosa* isolated from clinical and environmental samples in Minia, Egypt: prevalence, antibiogram and resistance mechanisms. *J Antimicrob Chemother 2007;60:1010-7.*

74. **Gill MM, Usman J, Kaleem F, Hassan A, Khalid A, Anjum R, et al.** Frequency and antibiogram of multi-drug resistant. *J Coll Physicians Surg Pak 2011;21:531-4.*

75. **Ndip RN, Dilonga HM, Ndip LM, Akoachere JFK, Nkuo Akenji T.** *Pseudomonas aeruginosa* isolates recovered from clinical and environmental samples in Buea, Cameroon: current status on biotyping and antibiogram. *Trop MedInt Health 2005;10:74-81.*

76. **Kpoda DS, Soubeiga AP, Karfo-Ouedraogo P, Ouedraogo OG, Gampene MT, Henry-Sangare R, et al.** Etude de la résistance aux antibiotiques des souches cliniques de *Pseudomonas aeruginosa*, isolees au laboratoire national de sante publique de Ouagadougou. *Sci Tech Sci Sante 2021;44:60-8.*

77. **Lob SH, Hoban DJ, Sahm DF, Badal RE.** Regional differences and trends in antimicrobial susceptibility of *Acinetobacter baumannii. Int J Antimicrob Agents 2016;47:317-23.*

78. **Mellouli A, Maamar B, Bouzakoura F, Messadi AA, Thabet L.** Colonisation and infection with *Acinetobacter baumannii* in a burns resuscitation unit in Tunisia. *Ann Burns Fire Disasters 2021;34:218-25.*

79. **Vacaroiu IA, Cuiban E, Geavlete BF, Gheorghita V, David C, Ene CV, et al.** Chronic kidney disease - an underestimated risk factor for antimicrobial resistance in patients with urinary tract infections. *Biomedicines 2022;10:2368.*

80. **Majeed HT, Aljanaby AA.** Antibiotic susceptibility patterns and prevalence of some extended spectrum beta-lactamases genes in Gram-negative bacteria isolated from patients infected with urinary tract infections in Al-Najaf City, Iraq. *Avicenna J Med Biotechnol 2019;11:192-201.*

81. **Pop-Vicas A, Strom J, Stanley K, D'Agata EM.** Multidrug-resistant gramnegative bacteria among patients who require chronic hemodialysis. *Clin J Am Soc Nephrol 2008;3:752-8.*

82. **Weiner LM, Webb AK, Limbago B, Dudeck MA, Patel J, Kallen AJ, et al.** Antimicrobial-resistant pathogens associated with healthcare-associated infections: summary of data reported to the national healthcare safety network at the centers for disease control and prevention, 2011-2014. *Infect Control Hosp*

Epidemiol 2016;37:1288-301.

83. **Lepape A, Machut A, Savey A.** Reseau national Rea-Raisin de surveillance des infections acquises en reanimation adulte - Methodes et principaux résultats. *Med Intensive Reanim 2018;27:197-203.*

84. **Centers for Disease Control and Prevention.** Antibiotic resistance threats in the United States, 2013. *Atlanta: CDC; 2013.*

85. **Weiner LM, Webb AK, Limbago B, Dudeck MA, Patel J, Kallen AJ, et al.** Antimicrobial-resistant pathogens associated with healthcare-associated infections: Summary of Data Reported to the National Healthcare Safety Network at the Centers for Disease Control and Prevention, 2011-2014. *Infect Control Hosp Epidemiol 2016;37:1288-301.*

86. **Zacharioudakis IM, Zervou FN, Ziakas PD, Mylonakis E.** Meta-analysis of methicillin-resistant *Staphylococcus aureus* colonization and risk of infection in dialysis patients. *J Am Soc Nephrol 2014;25:2131 -41.*

87. **Lu PL, Tsai JC, Chiu YW, Chang FY, Chen YW, Hsiao CF, et al.** Methicillin-resistant *Staphylococcus aureus* carriage, infection and transmission in dialysis patients, healthcare workers and their family members. *Nephrol Dial Transplant 2008;23:1659-65.*

88. **Nguyen DB, Lessa FC, Belflower R, Mu Y, Wise M, Nadle J, et al.** Invasive methicillin-resistant *Staphylococcus aureus* infections among patients on chronic dialysis in the United States, 2005-2011. *Clin Infect Dis 2013;57:1393-400.*

89. **Giarola LB, Dos Santos RR, Tognim MC, Borelli SD, Bedendo J.** Carriage frequency, phenotypic and genotypic characteristics of Staphylococcus aureus isolated from dialysis and kidney tranplant patients at a hosptial in northern Parana. *Braz J Microbiol 2012;43:923-30.*

90. **Zinai B, Mallek A** Etude epidemiologique de bacteries multiresistantes (BMR) cas des BLSE et SARM [Master]. *Oum El Bouaghi: Universite Larbi Ben M'hidi, Faculte des Sciences Exactes et des Sciences de La Nature et de la Vie; 2021.*

91. **Al-Orphaly M, Hadi HA, Eltayeb FK, Al-Hail H, Samuel BG, Sultan AA, et al.** Epidemiology of multidrug-resistant *Pseudomonas aeruginosa* in the Middle East and North Africa Region. *mSphere 2021;6:e00202-21.*

92. **Richet HM, Mohammed J, McDonald LC, Jarvis WR.** Building communication networks: international network for the study and prevention of emerging antimicrobial resistance. *Emerg Infect Dis 2001;7:319-22.*

93. **Cisneros JM, Rodriguez-Bano J.** Nosocomial bacteremia due to *Acinetobacter baumannii*: epidemiology, clinical features and treatment. *Clin Microbiol Infect 2002;8:687-93.*

94. **Higgins PG, Dammhayn C, Hackel M, Seifert H.** Global spread of carbapenem-resistant *Acinetobacter baumannii*. *J Antimicrob Chemother* *2010;65:233-8.*

95. **Murray BE.** The life and times of the *Enterococcus*. *Clin Microbiol Rev* *1990;3:46-65.*

96. **Wang TZ, Kodiyanplakkal RPL, Calfee DP.** Antimicrobial resistance in nephrology. *Nat Rev Nephrol 2019;15:463 -81.*

97. **Zacharioudakis IM, Zervou FN, Ziakas PD, Rice LB, Mylonakis E.** Vancomycin-resistant enterococci colonization among dialysis patients: a meta-analysis of prevalence, risk factors, and significance. *Am J Kidney Dis* *2015;65:88-97.*

98. **Nguyen DB, Shugart A, Lines C, Shah AB, Edwards J, Pollock D, et al.** National Healthcare Safety Network (NHSN) dialysis event surveillance report for 2014. *Clin J Am Soc Nephrol 2017;12:1139-46.*

99. 1. Broaders E, Gahan CGM, Marchesi JR. Mobile genetic elements of the human gastrointestinal tract: Potential for spread of antibiotic resistance genes. Gut Microbes. 12 Jul 2013;4(4):271-80.

100. Zhang T, Zhang XX, Ye L. Plasmid Metagenome Reveals High Levels of Antibiotic Resistance Genes and Mobile Genetic Elements in Activated Sludge. Gilbert JA, editor. PLoS ONE. 10 Oct 2011;6(10):e26041.

101. MONTAGNAC R, SCHILLINGER F, ELOY C. Prevention of bacteremia linked to central venous catheters in hemodialysis: benefit of treatment of the insertion site with a mixture of rifampicin and protamine. Nephrology (Geneva). 2003;24(4):159-65.

102. Pilly E. Maladies infectieuses et tropicales: prepa ECN, tous les items d'infectiologie. 6th ed. Paris: Alinea plus; 2019.

103. Gauzit R, Gutmann L, Brun-Buisson C, Jarlier V, Fantin B. Recommendations for the proper use of carbapenems. Antibiotiques. Dec 2010;12(4):183-9.

104. ISPD Peritonitis Recommendations: 2016 Update on Prevention and Treatment - Philip Kam-Tao Li, Cheuk Chun Szeto, Beth Piraino, Javier de Arteaga, Stanley Fan, Ana E. Figueiredo, Douglas N. Fish, Eric Goffin, Yong-Lim Kim, William Salzer, Dirk G. Struijk, Isaac Teitelbaum, David W. Johnson, 2016 [Internet]. [cite 12 Oct 2022]. Available from: https://journals.sagepub.com/doi/full/10.3747/pdi.2016.00078

105. Beaudreuil S, Hebibi H, Charpentier B, Durrbachr A. Serious infections in peritoneal dialysis and chronic conventional hemodialysis patients: peritonitis and vascular access infections. Reanimation. May 2008;17(3):233-41.

More
Books!

info@omniscriptum.com
www.omniscriptum.com
OMNIScriptum

Printed by Books on Demand GmbH, Norderstedt / Germany